Strategic Studies Institute
and
U.S. Army War College Press

AFRICA'S BOOMING OIL AND NATURAL GAS EXPLORATION AND PRODUCTION: NATIONAL SECURITY IMPLICATIONS FOR THE UNITED STATES AND CHINA

David E. Brown

December 2013

*****

FOREWORD

The Strategic Studies Institute (SSI) of the U.S. Army War College (USAWC) has maintained close and positive professional ties with our colleagues at the Africa Center for Strategic Studies (ACSS) in Washington, DC, since ACSS's founding in 1999. The Africa Center is the preeminent U.S. Department of Defense (DoD) institution for strategic security studies, research, and outreach in Africa.

I am pleased that SSI and ACSS are once more able to collaborate on this publication, the fourth SSI monograph authored by David E. Brown during his tenure as Senior Diplomatic Advisor at ACSS. Mr. Brown brings unique economic analysis and foreign policy perspectives to the important topic of this monograph: *Africa's Booming Oil and Natural Gas Exploration and Production: National Security Implications for the United States and China*. In September 2012, SSI published this author's monograph entitled *Hidden Dragon, Crouching Lion: How China's Advance in Africa is Underestimated and Africa's Potential Underappreciated*. In that monograph, the focus was on China's advance in Africa, while Africa's development was a sub-theme. This monograph reverses this emphasis, with the primary focus being Africa, specifically its oil and natural gas exploration and production, while China is a sub-theme, specifically the investment of its national oil companies (NOCs) in Africa and the implications of African energy developments for Beijing's national security and military strategy. Importantly, this present monograph is also different in that it includes a discussion of U.S. energy firms in Africa and the importance of the continent to America's energy security and national interests.

This monograph on Africa's energy future describes how the frenetic search for hydrocarbons in Africa has become so intense and wide ranging that there is planned or ongoing oil and gas exploration in at least 51 of the continent's 54 countries. Knowledge about Africa's geology is improving rapidly, generating great optimism about Africa's future growth and strategic position in the global economy. Because of a domestic boom in shale oil and gas in the United States, however, our nation's energy imports from Africa have been falling rapidly in recent years, raising the key strategic issue of whether Africa matters as much to U.S. energy security as it once did. Around 2030, North America is forecast to become a net oil exporter and will have little dependence on Africa for its energy needs. As Mr. Brown points out, the answer may be that, while Africa may be becoming less important for U.S. energy security, it is becoming more important for broader U.S. national security. He shows how this is so for a variety of reasons, such as the extraordinary trade and investment opportunities that this rapidly growing continent represents, including the need for $2.1 trillion in oil and gas sector investments between now and 2035 to realize its potential. Already, U.S. investments in African oil and gas in at least 22 countries on the continent are enormous, with prospects for future rapid growth.

At the same time, the United States is also an exception. As this monograph explains, for most major consuming regions, notably our Western allies and China, Africa has already become a crucial element in their energy diversification strategies. Africa is playing an increasing—not decreasing—role for them as an energy supplier. Africa's importance for Beijing's energy security continues to rise because China, despite shale discoveries of its own, will increasingly depend on energy imports until well past 2040. Examining also China's military white paper, Mr. Brown makes the key strategic prediction that, as superpower rivals in the 21st century, Washington's increasing energy security will make Beijing's own increasing energy insecurity be felt even more acutely, pushing the People's Liberation Army to accelerate adoption of a "two ocean" military strategy that includes an enduring presence in the Indian Ocean as well as in the Pacific Ocean.

At the end of this monograph, taking up almost two-thirds of its length, is an appendix with a series of country profiles on oil and gas exploration and production in 51 of 54 African nations and the disputed territory of the Western Sahara. The profiles focus on U.S. and Chinese energy companies, but also discuss those from other developed and emerging nations.

SSI is pleased to offer this monograph to assist U.S. Army and DoD senior leaders and strategic thinkers in understanding the key issues of the day.

DOUGLAS C. LOVELACE, JR.
Director
Strategic Studies Institute and
    U.S. Army War College Press

ABOUT THE AUTHOR

DAVID E. BROWN is a career member of the Senior Foreign Service, who joined the Africa Center for Strategic Studies (ACSS) as Senior Diplomatic Advisor in August 2011. His prior African experience includes serving as the Senior Advisor to the J-5 (Strategy, Plans, and Programs) Director of the U.S. Africa Command (AFRICOM) in Stuttgart, Germany; three times as Deputy Chief of Mission at U.S. Embassies in Cotonou, Benin; Nouakchott, Mauritania; and Ouagadougou, Burkina Faso; and as Economic Officer at the U.S. Consulate-General in Lubumbashi, Democratic Republic of the Congo. Mr. Brown's non-Africa overseas tours have been as Consul General in Chengdu, China; and Economic Officer in Beijing; Tokyo, Japan; and Moscow, Russia. He has also served in Washington, DC, as the Director of the Office of Environmental Policy; as Economic Officer in the Bureau of Economic and Business Affairs responsible for trade policy with developing countries, including Africa; and on the Canada desk, with responsibilities for economic, consular, and law enforcement issues. Prior to joining the U.S. Department of State, he worked in Miami as the business manager of the Latin American Bureau of CBS News. Mr. Brown holds a B.A. in government (political science) from Cornell University; an MBA from the University of Chicago, specializing in finance; and an MBA from the University of Louvain, Belgium, with majors in econometrics and international business.

SUMMARY

The frenetic search for hydrocarbons in Africa has become so intense and wide ranging that there is planned or ongoing oil and gas exploration in at least 51 of the continent's 54 countries. Knowledge about Africa's geology is improving rapidly, generating great optimism about the continent's energy future. Onshore and offshore rifts and basins created when the African continent separated from the Americas and Eurasia 150 million years ago are now recognized as some of the most promising hydrocarbon provinces in the world. Offshore Angola and Brazil, Namibia and Brazil, Ghana and French Guyana, Morocco and Mexico, Somalia and Yemen, and Mozambique and Madagascar are just a few of the geological analogues where large oil fields have been discovered or are believed to lie. One optimistic but quite credible scenario is that future discoveries in Africa will be around five times their current level based on what remains unexplored on the continent versus currently known sub-soil assets. If proven true, this could have a profoundly positive impact on Africa's future growth and strategic position in the global economy.

Africa had proven oil reserves of 132.4 billion barrels at the end of 2011, an increase of 154 percent over the 1980 figure of 53.4 billion barrels. Because of definitional issues of what constitutes "proven reserves," however, this figure likely grossly underestimates Africa's oil and gas potential and does not include likely future reserves in the Mauritania Basin in the country of the same name (discovered in 2001); the Albertine Basin in Uganda, which straddles the Democratic Republic of the Condo (DRC) and the Kenyan part of the East African Rift (2006 for Uganda, 2012 for Kenya);

the Tano Basin in Ghana (2007); the Rovuma Basin in Mozambique and Tanzania (2010 in both countries); and the Sierra Leone/Liberia Basin in both countries (2010 for Sierra Leone, 2012 for Liberia). Moreover, no unconventional reserves have been added to these "proven reserve" estimates despite the fact that the continent has substantial, proven heavy oil/bitumen in several countries including Congo (Brazzaville), Nigeria, and Madagascar, and potential shale gas, most notably in South Africa, Algeria, Libya, Tunisia, and Ethiopia.

Africa has already become a crucial element in the energy diversification strategies being adopted in major consuming regions, notably China, and offers relative freedom for international oil companies (IOCs) to invest and operate. Africa will be the world's fourth most important region for **oil** production from 2010 to 2035, after the Middle East and Europe/Eurasia, not far behind North America, and ahead of Latin America and East Asia/Australia. Africa is also starting to play a more prominent role in international markets for **natural gas,** which is currently characterized by disparate regional markets but is slowly moving toward a more interconnected, global gas market. One reason for this is the Panama Canal expansion, which will allow much larger PANAMAX LNG tankers to unite the Atlantic Basin and the Pacific Basin—much as the advent of oil supertankers fostered a global market in oil. Africa's total natural gas production will increase rapidly over the next 2 decades, from 188.1 million tons of oil equivalent (TOE) in 2010 to 257.2 million TOE in 2020 and 356.8 million TOE in 2030.

Because of a domestic boom in shale oil and gas in the United States, our nation's energy imports from Africa have been falling rapidly in recent years, raising the key strategic issue of whether Africa matters as much to U.S. energy security as it once did. Around 2030, North America is forecast to become a net oil exporter—an outcome that also means the United States will enjoy an order of magnitude greater energy security in 20 years than it does at present, and will have little dependence on Africa for its energy needs. The answer may be that, while Africa may be becoming less important for U.S. energy security, it is becoming more important for broader U.S. national security. This is so for a variety of reasons, such as the extraordinary trade and investment opportunities that this rapidly growing continent represents, including the need for $2.1 trillion in oil and gas sector investments between now and 2035 to realize its potential. Already, U.S. investments in African oil and gas in at least 22 countries on the continent are enormous, with prospects for future rapid growth.

Africa is also playing an increasing—not decreasing—role as an energy supplier to our Western allies and China. Africa's natural resources, above all oil and gas, are already China's top national security interest in the continent and will remain so in the coming decades. In December 2012, China overtook the United States as the world's largest oil importer. Africa's importance for Beijing's energy security continues to rise because China, despite shale discoveries of its own, will increasingly depend on energy imports until well past 2040. As superpower rivals in the 21st century, Washington's increasing energy security will make Beijing's own increasing energy insecurity be felt even more acutely, pushing the People's Liberation Army to accelerate adoption of a "two ocean" military strategy that includes an enduring presence in the Indian Ocean as well as the Pacific Ocean.

Up to now, China's "oil diplomacy" in Africa has been most successful in niche markets such as Angola (due to corruption and to its $14 billion in infrastructure-for-oil deals), and in other African countries, such as Sudan and Niger, with fields that have been ignored as politically sensitive or too marginal by major Western companies. Chinese oil companies are rapidly increasing their energy investments in several other African nations, however, and already have a significant presence in three major oil and gas producing countries — Algeria, Angola, and Nigeria — and in four smaller producers — Cameroon, Chad, Gabon, and Sudan.

At the end of this monograph, taking up almost two-thirds of its length, is an appendix with a series of country profiles on oil and gas exploration and production in 51 of 54 African nations and the disputed territory of the Western Sahara. The profiles focus on U.S. and Chinese energy companies and discuss those from other developed and emerging nations.

GLOSSARY

bbbl	billion barrels of oil
bbl	oil barrel
bcf/d	billion cubic feet per day
bcf/y	billion cubic feet per year
bcm	billion cubic meters
BOE	barrels of oil equivalent
BOE/D	barrels of oil equivalent per day
bn	billion
bpd	barrels per day
btu	British thermal unit
cf/d	cubic feet per day
EEZ	Exclusive Economic Zone
EIA	U.S. Energy Information Administration
EITI	Extractive Industries Transparency Initiative
EOR	enhanced oil recovery
E&P	exploration and production
GIP	gas in place
IEA	International Energy Agency
IOC	International oil company
JDZ	Joint Development Zone
LNG	liquefied natural gas
mmscf/d	million standard cubic feet per day
mmt/y	million metric tons per year
mt/y	metric tons per year
mpta	million tons per annum
m3/d	cubic meters per day
NOC	National Oil Company
OECD	Organisation for Economic Co-operation and Development
OOIP	original oil in place
OPEC	Organization of Petroleum Exporting Countries

PSA	Production Sharing Agreement
PSC	Production Sharing Contract
TOE	tons of oil equivalent
tcf	trillion cubic feet
tcm	trillion cubic meters
UN	United Nations
USGS	United States Geological Survey

AFRICA'S BOOMING OIL AND NATURAL GAS EXPLORATION AND PRODUCTION: NATIONAL SECURITY IMPLICATIONS FOR THE UNITED STATES AND CHINA

INTRODUCTION

In September 2012, the Strategic Studies Institute (SSI) of the U.S. Army War College (USAWC) published this author's monograph entitled *Hidden Dragon, Crouching Lion: How China's Advance in Africa is Underestimated and Africa's Potential Underappreciated*. In that monograph, the focus was on China's advance in Africa, while Africa's development was a sub-theme. This monograph reverses this emphasis, with the primary focus being Africa, specifically its oil and natural gas exploration and production (E&P), while China is a sub-theme, specifically the investment of its national oil companies (NOCs) in Africa and the implications of African energy developments for Beijing's national security and military strategy. Importantly, this present monograph is also different in that it includes a discussion of U.S. energy firms in Africa and the importance of the continent to America's energy security and national interests.

Reasons for Optimism about Africa Oil and Gas Potential.

About 75 percent of the undiscovered technically recoverable oil outside of the United States is in four other areas of the globe: Africa, the Middle East, South America and the Caribbean, and the Arctic provinces of North America.[1] The author has intentionally listed Africa first among these regions—even ahead of the

Middle East—because its enormous long-term potential is significantly underestimated in conventional analyses, which do not look beyond 2040.[2] Why such optimism about Africa's long-term leadership role in the global energy market?

First, the research for this paper, based on over 1,000 source documents, revealed an extraordinary picture of momentous changes in Africa, both in understanding of the continent's geology and of the frenetic pace of ongoing oil and gas exploration. The hunt for hydrocarbons in Africa has become so intense and wide ranging that the author could find no evidence of planned or ongoing oil and gas exploration in only three of the continent's 54 countries: Burkina Faso, a smaller country in inland West Africa; and Lesotho and Swaziland, two small nations that are, respectively, wholly or partially surrounded by the Republic of South Africa. With few exceptions—such as established producers Algeria and Nigeria, where short-sighted government policies have discouraged Western oil producers from expanding investment—Africa has been the site of a veritable exploration boom. In country after country, one reads of how a well has been drilled for the first time in "X" number of years, sometimes well over 20 years. Even the continent's "ultra-frontier" countries are being explored. Just one example is the extensive, ongoing exploration efforts in troubled Somalia—a country whose central government dissolved into chaos in 1991 and is still plagued by terrorist group Al Shabaab. This exploration boom in Africa has already led to a series of new oil and gas fields and is laying the foundation for future large discoveries. The 2007 discovery of oil in Ghana in West Africa, for example, has led oil companies to pursue offshore drilling all along the continent's Atlantic coast. The U.S. Geological Survey (USGS) has estimated that West Africa alone also contains 120 billion barrels of oil equivalent (BOE) of undiscovered petroleum resources.

Second, despite this accelerating oil and gas exploration, the continent is enormous and remains largely underexplored for hydrocarbons. While Africa today holds 8 percent of proven global oil and gas reserves, it accounts for 20.4 percent of the world's land mass (and 22.4 percent when Antarctica is excluded).[3] It also has a coastline of 16,000 miles (26,000 kilometers [km]) with Exclusive Economic Zones (EEZs) extending 200 miles offshore[4] and seven island nations with their own large EEZs.[5] There were only 1,000 wells drilled onshore and offshore in Africa in 2011 compared with 18,500 drilled in Alberta, Canada, in 2005.[6] While cumulatively, some 20,000 wells have been sunk in North Africa[7] and 15,000 wells have been drilled in West Africa, only 500 have been drilled to date in East Africa.[8] Even with the modest number of wells drilled in East Africa, the last 5 years have seen the discovery of major oil and gas reserves in Uganda, Kenya, Mozambique, and Tanzania—proving that several of Africa's remaining unexplored sedimentary basins promise to contribute an even larger share of global production and reserves in the future.[9] One less conventional view supported by a recent Ernst and Young report and scholar Paul Collier is that the scale of what is likely to happen in Africa is not widely appreciated and that:

> future discoveries and resulting exports of resources (including oil and gas) will be around five times their current level (based on what remains unexplored in Africa versus currently known sub-soil assets), and that this will have a profoundly positive impact on

Africa's future growth and strategic positioning in the global economy.[10]

Good Governance in the African Energy Sector and Avoidance of the Resource Curse.

While this monograph is primarily about oil and gas exploration and production, the author will take a moment to stress that good governance in Africa's oil and gas industry will play a critical role—indeed arguably the **central** role—in how much the continent and its people will benefit from this bonanza of energy wealth.[11] In this regard, the author's home institution, the Africa Center for Strategic Studies, held a conference in Abuja, Nigeria, in 2005 about the interplay between Africa's energy future and human security. The key conclusions of this conference are still valid:

> Increased [oil and gas] wealth without adequate constitutional, legal, regulatory, and policy reforms could exacerbate human security challenges [in Africa and] include an overreliance on oil and gas revenues, weak political and economic governance, corruption, relative poverty, and resource tensions. The complex linkages among human security, weak governance, and the oil and gas sectors highlight the primacy of internal threats in African countries including community grievances, sabotage, theft, and militant groups.[12]

This monograph discusses good governance mainly in the context of the need to attract foreign investment and reduce risk. While its central focus is not on good governance and human security, this is only because of the practical need to limit the scope of this paper to the already expansive issue of African oil and gas exploration and production.

I. IN THE BEGINNING: AFRICAN OIL AND CONTINENTAL DRIFT THEORY

Perhaps the most important geological **concept** for the layman to grasp Africa's bright future in oil and gas is the theory of **continental drift**. In paleogeography, the study of historical geography, Gondwana is the name given to the more southerly of two supercontinents (the other being Laurasia) that existed from approximately 180 to 510 million years ago. Gondwana, also called Gondwanaland, incorporated present-day South America, Africa, Madagascar, Saudi Arabia, India, Australia, and Antarctica. Under this theory, the waters off much of the Americas today are a mirror image of the rich oil deposits off the western side of Africa, and vice versa. Geological evidence suggests, for example, that Morocco, where there is now feverish offshore exploration, was attached to present-day Mexico, whose rich petroleum resources have been known for decades; Ghana, where the huge Jubilee oil field was discovered in 2007, was attached to present-day French Guyana, where an over one billion-barrel field was discovered in 2011; and Angola, where the first offshore field was discovered in 1966, was attached to Brazil, where exploration began in 1968, but the huge offshore oil fields at Campos were discovered more recently.[13]

Two geological vocabulary terms that are important to understanding African oil and gas exploration are **rifts** and **grabens**. A rift is a linear zone where the Earth's crust and its outermost shell, or lithosphere,[14] are being pulled apart. Rifts are important because the sedimentary rocks associated with continental rifts often host important deposits of hydrocarbons.[15] A graben is a depressed block of land bordered by parallel faults.[16] Grabens are also commonly associated with the presence of hydrocarbons. The East Africa Rift Valley, for example, is rich in grabens and, not coincidentally, is also increasingly recognized in the oil industry as a highly promising "petroleum province." A third geological vocabulary term important for Africa is what is called a **pre-salt layer**, which is a geological formation regularly found on the continental shelves off the coast of Africa and the Americas. Pre-salt layers are important because they often hold significant petrochemical resources. For the Atlantic, oil and natural gas tend to lie below an approximately 2,000-meter (m) deep layer of salt, itself below an approximately 2,000-m deep layer of rock under 2,000-3,000-m of water.[17]

Much of the oil and gas found offshore in Africa and the Americas has been in pre-salt layers, for which seismic imaging has improved considerably in recent years. The following statement from U.S. firm Cobalt International captures the essence of the importance of continental drift theory for the entire western or Atlantic side of Africa from Morocco to South Africa:

> The pre-salt geology of offshore West Africa has many similarities to that of offshore Brazil, the location of a number of recent super-giant discoveries. Existing on-shore and near-shore pre-salt fields in West Africa provide additional support for the existence of deep-water pre-salt opportunities. Over the past 150 million years, rifting and separation of South America from Africa has occurred along much of the Atlantic Margin. The pre-rift geology below the salt [layer] remains largely unaltered and is consistent from one continent to the other. This has resulted in a series of analogous

basins on both sides of the Atlantic. Each of these basins contains a world-class oil-prone source rock and an extensive salt layer, which acts as a regionally extensive seal.[18]

The eastern, or Red Sea/Indian Ocean, side of Africa has a similar geological story, albeit more complex. The East African Rift System is the largest continental rift system on the Earth's surface and actually extends from the Jordan in the Middle East to Mozambique in southern Africa. Within the East African Rift System are two branches of note: the East Rift Valley (often called the Great Rift Valley) and the Western Rift Valley, both in central-eastern Africa. The East Rift Valley System is increasingly felt to also include a third, southwesterly branch ending in Namibia. The northern part of East Africa was also part of the Gondwana mega-continent and drifted apart from the modern day Arabian Peninsula. Since the Red Sea Basin has had several large oil and gas discoveries in Egypt, Sudan, and Saudi Arabia, oil and gas prospectors have recently noted with great excitement the geological similarities between the Darin and Nogul Basins in Somalia and the Masila and Marib-Shabwa Basins in oil-rich Yemen. Anglo-Irish Tullow Oil used the discovery of oil in Uganda in 2006 to look for rift valley analogies in Kenya and Ethiopia. This same theory of continental drift has also been applied to oil and gas found along the coast of Madagascar since that island nation shares similar geological patterns with Kenya, Tanzania, and Mozambique, to which it was connected 145 million years ago. The Seychelles is also thought to have hydrocarbon potential because it can also be reconstructed to a position adjacent to Somalia and as a northern extension of Madagascar.

II. A GEOLOGICAL MAP OF AFRICA

Most readers interested in Africa have some familiarity with the political division of the continent into regional economic communities in West, Central, South, East, and North Africa, such as the Economic Community of West African States (ECOWAS). This monograph, however, eschews the traditional political map of Africa in favor of a geological map. The author has chosen this approach because he is convinced that a geological map will enable readers to more clearly understand and visualize both the interconnected nature of oil and gas fields straddling countries, and the political imperative in many cases for future cross-border cooperation in jointly developing related resources and infrastructure.

Major Country-level Trends in Africa's Geological Sub-Regions.

The six geological sub-regions of Africa presented here are:

1. The North African Margin, geologically includes northern Morocco roughly east of the Strait of Gibraltar, which is part of the North Africa Margin.[19] It also includes Algeria, Tunisia, Libya, and Egypt.

2. The Atlantic Transform Margin[20] runs geologically via from Morocco roughly west of the Strait of Gibraltar, all along the western coast of Africa to western South Africa. Also included are the countries of Mauritania, Senegal, Gambia, Cape Verde, Guinea Bissau, Guinea, Sierra Leone, Liberia, Cote d'Ivoire, Ghana, Togo, Benin, Nigeria, Sao Tome and Principe, Cameroon, Equatorial Guinea, Gabon, Congo (Braz-

zaville), the Democratic Republic of the Congo (DRC), Angola, and western Namibia. (The former Spanish colony and currently disputed territory of Western Sahara is also included in this sub-region.)

3. The East Rift Coastal[21] is a branch of the larger East African Rift System and runs geologically from the Red Sea coast of Egypt, Sudan, Eritrea, and Djibouti southward to Somalia, Ethiopia, eastern Kenya, eastern Tanzania, eastern Mozambique, western Madagascar, eastern South Africa, and the island nations of the Comoros and Seychelles. (Juan de Nova, which is part of the French Southern and Antarctic Lands, is discussed jointly with Madagascar;[22] Mauritius is at best distantly related to this System, but is grouped here for the sake of simplicity.)

4. The West Rift Valley, a second branch of the East African Rift System, runs geologically from southeastern Sudan, the country of South Sudan, Uganda, eastern DRC, Rwanda, Burundi, western Tanzania, Zambia, Malawi, and western Mozambique.

5. The Central Africa Rift System[23] includes a series of basins running through northern Cameroon, southern Chad, and northern Central African Republic, intersecting basins such as Jongolei and Melut that also extend into South Sudan. Three other countries in inland West and Central Africa do not fit our model, but are included in this section for the sake of simplicity: Mali, which has basins extending into its neighbors, Mauritania, Algeria, and Niger; Niger, whose basins extend into Mali, Algeria, Libya, Chad, and Nigeria; and Burkina Faso, which has no known hydrocarbons.

6. Southern Africa is at the end of both the Atlantic Transform Margin in the west to the East African Rift in the east. The Southern Africa geological region also has such complexities such as the Karoo Basin shale

oil in its west and coal-bed methane in its northeast. Included here are eastern Namibia, Botswana, South Africa, Zambia, Zimbabwe, and eastern Mozambique. There are no known hydrocarbons in Swaziland and Lesotho.

For these six geological regions, the author summarizes here the highlights of information found in the country profiles in Appendix 1.

1. *The North Africa Margin – Historical Leader in Oil Production, but Shaken by Political Instability.*
Geologically speaking, the North Africa Margin could be taken roughly to include Algeria, Tunisia, Libya, and Egypt. Cumulatively, the countries of North Africa, here excluding Morocco, have historically been the single most important oil and gas region in Africa. In 2010, Libya, Algeria, and Egypt were, respectively, the third, fourth, and fifth largest oil producers in Africa after Nigeria and Angola. Algeria, Egypt, and Libya were even more dominant in natural gas production, occupying the number one, two, and four positions in 2009 (with Nigeria third and Equatorial Guinea fifth). North Africa has been primarily an onshore "conventional play," but increased offshore exploration across the sub-region (including eastern Morocco) promises significant new discoveries. The sub-region has broadly remained open to IOCs, with decades-long history of activity by both foreign investors and NOCs, but with bouts of resource nationalism that have periodically dampened the enthusiasm of foreign investors.

At present, North Africa is beset with political and economic/demographic challenges. Politically, the Arab Spring set off regime changes in Tunisia, Egypt, and Libya that have made the sub-region more democratic, but less stable in the short-term. The current political uncertainty in North Africa, including the terrorist attack by al-Qaeda in the Islamic Maghreb (AQIM) at the Amenas gas project in Algeria in January 2013 and ongoing, low-intensity fighting between militia groups in Libya, is having significant implications for the region's oil and gas industry, both in terms of disrupted operations and postponed investment. The decision by U.S. firm ConocoPhillips in December 2012 to sell its Algerian assets, and the decisions by Anglo-Dutch Shell and U.S. firm Chevron to abandon drilling and exploration in several blocks in Libya, reflect deep concern about the risk-return calculus in North Africa. Algeria, Tunisia, Libya, and Egypt face similar economic/demographic challenges that have also caused domestic demand for oil and gas to soar: heavily subsidized fuel prices, rapidly rising populations, and greatly increasing car use. This increased demand for oil and gas has, in turn, caused the amount available for export to decline, thereby also sharply reducing government revenues.

In Algeria, the unfavorable 2005 hydrocarbon law and a 2006 presidential decree have dampened the enthusiasm among IOCs for investment in the country's oil and gas sector, contributing to a 20 percent drop in hydrocarbon production over the last 5 years. In January 2013, Algeria's parliament approved amendments to the law, but some observers remain pessimistic that these changes will turn around declining foreign investment. One apocalyptic analysis even predicted that Algeria could run out of available oil to export after 2023 because of soaring domestic demand and falling production.

In Libya, political volatility and a tenuous security situation following the 2011 civil war have also dampened foreign investment, which had been on the resurgence in the mid-2000s after the United Nations (UN) and United States lifted sanctions. For the next few years, output in Libya is expected to stay near the 1.7 million barrels per day (bpd) produced in 2010. After this, output may resume a slow upward path if new fields come into production to replace existing fields entering their decline phases. One reason for longer-term optimism about Libya's energy export potential is that, with proven oil reserves of 47.1 billion barrels, the largest in Africa, the country is simply too important to ignore.

Post-Arab Spring, the oil and gas sectors of Tunisia and Egypt have fared better. Tunisia had a modest upstream industry to begin with, with declining production for much of the last 20 years and no exports, but with new exploration that could potentially reverse this decline. Political unrest has not directly affected foreign oil investors in Egypt, as the top producing regions — the offshore Nile Delta and the Western Desert — are far from population centers. Overall, Egyptian oil production has been declining, but new discoveries have continued and promise hope to reverse this decline. Since 2000, oil output in the Western Desert has doubled, for example, and now accounts for around 28-30 percent of Egypt's total oil production.

2. *Atlantic Transform Margin – Common Offshore Geology with Latin America Highly Promising.*

The Atlantic Transform Margin runs from Morocco roughly west of the Strait of Gibraltar, all along the western coast of Africa via the Gulf of Guinea, to western South Africa. (The DRC is addressed later in the section on the Central Africa Rift System.)

This geological sub-region is dominated by one of the old lions (Nigeria) and one of the new lions (Angola). The outlook for Organization of Petroleum Exporting Countries (OPEC) member Nigeria, Africa's largest oil producer, hinges on resolving a number of major issues, including civil conflict in producing regions; improving the transportation infrastructure, whose deteriorated state has led to sporadic disruptions to supply; the inability of the Nigerian National Petroleum Company (NNPC) to fund its share of investment in joint ventures with international companies; and delays in agreeing to a new Petroleum Industry Bill, which has paralyzed investment in recent years.[24] ConocoPhillips' pessimistic view of Nigeria's investment environment was behind the $1.79 billion sale in January 2013 of its Nigeria operations.

Although crude oil production in Angola, Africa's second largest oil producer and an emerging major gas supplier, slipped to 1.79 million barrels per day (bpd) in 2011, the government is targeting 2 million bpd production levels by 2014. Successful exploration in Angola's pre-salt formations continues to drive optimistic oil production forecasts for the country and hopes that new deepwater discoveries will yield more than enough oil to offset declines at existing fields.[25]

South-Central Africa: Namibia to Sao Tome and Principe. South of Angola, Namibia attracted investment by major IOCs in 2012 on the bet that its coastal shelf may mirror that of Brazil across the Atlantic Ocean, where the Tupi discovery in 2007 was the biggest find in the Americas in 3 decades.[26] North of Angola, Congo (Brazzaville) was still Africa's seventh largest oil producer in 2010, although its oil production is in decline and potential gas development is limited by a lack of infrastructure. Oil and gas production could increase, however, if the country's deepwater offshore potential[27] is realized. Congo also has a pilot oil sands project and has agreed to jointly exploit the Lianzi cross-border oil field with Angola.

Crude oil production in Equatorial Guinea, Africa's eighth largest producer in 2010, has been declining as a result of maturing oil fields, but with the recent start-up of the Aseng oil and gas-condensate field and the anticipated start-up of the Alen gas-condensate field in late-2013, the development of these new fields are projected to revive liquids production in the near term.[28] Oil output in Gabon has dropped about 40 percent from its peak of 370,000 bpd in 1997, causing Gabon to fall to the ninth largest oil producer in Africa.[29] In the long run, Gabon's oil production will depend on the success of new exploration, particularly of deepwater, pre-salt fields, interest in which has been piqued by recent offshore Brazil pre-salt discoveries.[30]

While Cameroon's oil production has been declining since 1985,[31] the government announced that it expected production would rise about 59 percent by the end of 2012 because of the development of new fields. There have been significant recent discoveries of oil and gas condensates in Cameroon's two main basins.[32] The northern Bakassi region is considered to be rich in oil and gas, but had little exploration until recently due to lingering concerns about a territorial dispute with Nigeria, which was adjudicated in Cameroon's favor in the mid-2000s.[33] To date, there have not been any major commercial oil or gas finds in Sao Tome and Principe, but there is guarded optimism about future discoveries. Three oil and gas stories for Sao Tome and Principe concern 1) exploration and production (E&P) in the Joint Development Zone (JDZ) with Nigeria; 2) E&P in Sao Tome's EEZ; and 3) plans to build a tank farm and oil transshipment port for international use firms operating in the Gulf of Guinea sub-region.

Ghana Discoveries Bring Hope from Benin to Morocco. The 2007 discovery of Ghana's Jubilee Field is transforming the country's oil and gas industry and estimations of oil and gas potential all along Africa's Atlantic coast. Jubilee and other successes in adjacent fields bode particularly well for neighboring offshore developments in the Atlantic Transform Margin from Benin and Togo to the west, to Cote d'Ivoire, Liberia, Sierra Leone, and Guinea to the east.[34] To date, Benin had limited production at its Seme offshore field, which may reopen in the future. In June 2012, oil was discovered in the Dahomey Basin in Togo, which is part of the Atlantic Transform Margin extending from the Keta Basin in eastern offshore Ghana eastward through Togo and Benin to the Mont Okitipupa Basin in Nigeria.[35]

Cote d'Ivoire has proven reserves of only 100 million barrels of oil and 1.1 trillion cubic feet (tcf) of natural gas, but oil and gas finds in neighboring Ghana are spurring additional exploration and have already led to a number of new discoveries since 2010. In Liberia, a new offshore oil discovery in February 2013 boosted the already considerable optimism about the possibility of a new offshore hydrocarbon sub-province known as the West Africa Transform Margin. Recent oil discoveries in Sierra Leone confirmed that the main reservoir and source elements in the Jubilee field in Ghana are present in the Sierra Leone/Liberia deepwater basin.[36] Exploration remains at an early stage, but prospects look good for Sierra Leone to become a new frontier for oil and gas.[37]

In Guinea, IOCs are teaming up to "apply their successful exploration experience on the Atlantic

Margin . . . particularly the [West Africa] Transform Margin play that is present in the Guinea acreage."[38] There have been recent oil discoveries in both Guinea-Bissau's offshore blocks and its JDZ with Senegal, but the country's reputation as a target for foreign investment has been severely damaged by perceptions that it had devolved into a narco-state under the influence of Latin American drug cartels.[39]

The main oil and gas E&P areas in Senegal are in offshore blocks in Senegal's territorial waters or EEZ, and in Senegal's JDZ with Guinea-Bissau. As with Mauritania, in pre-rift time, Senegal was considered to be adjacent to Mexico before the African and American continents pulled apart.[40] Tullow is optimistic about Senegal because an exploratory well drilled in 2011 had the same cretaceous turbidite play that led Tullow to discover the Jubilee Field in Ghana. There are no known oil or gas reserves in the Gambia, but the country has good prospectivity for hydrocarbons since the area marks the northern extent of the Casamance-Bissau sub-basin, which forms part of the Mauritania-Senegal-Gambia-Guinea Bissau Coast Basin,[41] also known in the industry as the "MSGBC" Basin, which itself is part of the geological sub-region known as the North-West Africa Atlantic Margin.

Mauritania is a frontier exploration country where the limited oil and gas production has been disappointing so far, but where promising offshore and onshore prospects have attracted Asian, Australian, European, and U.S. oil firms. Mauritania has the same geological source rocks as found in Ghanaian oil fields.[42] Some experts have suggested that oil and gas reserves discovered in neighboring Mauritania could stretch as far as Cape Verde, but there has been only limited exploration so far in this island nation and its huge EEZ.

Compared to its prolific eastern neighbor Algeria, Morocco has had a modest upstream industry. However, most sedimentary basins in Morocco onshore and offshore are still largely unexplored,[43] with IOCs attracted to the country's considerable potential and the government's favorable tax regime. Both Morocco and the Saharawi Arab Democratic Republic (SADR)[44] have granted exploration licenses to IOCs in Western Sahara despite a UN resolution against such deals until the issue of the territory's sovereignty has been resolved. However, no OICs have been carrying out drilling in this disputed territory.

3. *East Rift Coastal Region – Geological Links to Oil-Rich Saudi Peninsula, Madagascar.*

The East Rift Coastal region is a branch of the East African Rift System, which runs geologically on the continent from the Red Sea coast of Egypt to eastern South Africa.

Sudan/South Sudan: Continued Conflict, but Bound Together by Pipeline. After a decade of war, Sudan and South Sudan signed a peace agreement in 2005 after a referendum that led to the independence of South Sudan in July 2011. Sudan and South Sudan, which, when aggregated together were the sixth largest oil producer in Africa in 2010, saw average daily production in 2011 fall to 425,000 barrels, a significant decline from 470,000 bpd in January 2011. A dispute over pipeline and export transit fees lingered post-independence, leading South Sudan to halt exports for several months, until March 2013. South Sudan is eager to export its oil via Kenya to reduce its dependence on Sudan.

Eritrea, Djibouti, and Somalia: Geological Links to Oil-Rich Yemen. Eritrea's great potential over the long term as an oil and gas producer—with geological analogues both in Uganda's rich petroleum basin and with Yemen across the Red Sea—has not been realized, in part due to political instability, including wars with Ethiopia and an autocratic regime that was subjected recently to UN sanctions that are still ongoing. The Red Sea Basin, of which Djibouti is part, has had several discoveries in its northern area in Sudan, Egypt, and Saudi Arabia, and there are also analogues with the large Jurassic oil fields located nearby in Yemen.[45] Djibouti's Guban Basin extends into Somalia's northern province of Somaliland, where there have been oil seeps and recovery of oil from wells drilled in previous decades.

Estimates of onshore and offshore reserves of Somalia vary greatly, but run as high as 110 billion barrels of oil[46]—a staggering figure that, if proven true, would give this country the continent's largest reserves. Independent international E&P companies are presently investing in Puntland and Somaliland despite the legal and political uncertainties. Parts of Puntland are geological extensions of basins in oil-rich Yemen, while Somaliland contains geological basins similar to both Yemen and Uganda, where millions of barrels of oil reserves have been discovered.

Ethiopia and Kenya: Crossroads of Several Geological Basins. There are five sedimentary basins in Ethiopia with prospects for hydrocarbons. Some geologists believe that there are several tcf of gas condensate in Ogaden, which they see as just one of a large number of promising East African coastal basins. Ethiopia is also in part of the Turkana Rift Basin, shared with Kenya. This basin is similar in geology to the Lake Albert Rift Basin, where large amounts of oil have been discovered in Uganda. Ethiopia also has oil deposits on the border with Eritrea to the north and Sudan to the west.

Major oil discoveries in Uganda in 2006 and natural gas discoveries in Mozambique and Tanzania in 2010 have created expectations that Kenya is part of a new regional hydrocarbon province. Kenya has four large sedimentary basins offshore and onshore. These are found in southeast Kenya, which has similar geology to Madagascar, and in the Tertiary Rift Basin in western Kenya, which borders Sudan and Ethiopia to the north, Uganda to the east, and Tanzania to the south.[47] In 2012, discoveries were made of oil in the Lochikar onshore sub-basin[48] and of natural gas offshore.[49]

Tanzania and Mozambique: A New Hydrocarbon Frontier. Recent, large discoveries of offshore natural gas in Tanzania and Mozambique have led observers to dub these two countries as Africa's "newest lion cubs," and led to soaring expectations about East Africa's potential as a new hydrocarbon frontier. Recent discoveries in Tanzania have boosted its natural gas reserves to over 34 tcf, leading to hopes for the country's first multitrain liquefied natural gas (LNG) development.[50] In February 2013, the government announced that it would soon issue a license to an IOC to explore in Lake Tanganyika, an area that is also currently claimed by Malawi.

Mozambique has two main sedimentary basins: the Rovuma Basin in the northeast adjacent to the Tanzanian border and the Mozambique Basin in the south.[51] Rovuma Area 1 has up to 100 tcf of natural gas in place, making it among the world's most significant hydrocarbon finds in the last decade.[52] Rovuma Area 4 has an estimated 75 tcf of natural gas in place. Several oil companies are also carrying out exploration activities in the Urema Graben in central Mozambique, which is in the southern part of the East African Rift System.[53] In Mozambique, U.S. firm Anadarko and Italy's ENI are jointly working to advance an LNG park drawing upon Areas 1 and 4's gas to produce approximately 50 million tons of LNG per year—enough to elevate the country to one of the world's largest LNG exporters.[54]

Island Nations and Territory: Madagascar, Comoros, Seychelles, Mauritius, and Juan de Nova. The island nation of Madagascar has five sedimentary basins.[55] The onshore part of the Morondava Basin is a proven petroleum province with onshore discoveries of oil sands and subsurface heavy oil deposits exceeding 20 billion barrels. The offshore portion of the basin is considered to be part of the same petroleum system, but extending into a deeper geologic setting that may yield lighter oil discoveries.[56] Political instability has been a serious impediment to foreign investment in Madagascar's oil and gas industry and only made worse by threats by the transitional government to expropriate major oil fields.[57]

In the Mozambique Channel between Mozambique and Madagascar lie the Juan de Nova islands, which are controlled by France and where exploration is occurring east of the gas-rich Rovuma Basin in Mozambique. The island nation of the Comoros awarded its first oil exploration license in March 2012 for acreage adjacent to offshore Area 1 and Area 4 of Mozambique's Rovuma Delta. Global interest in the waters off the island nation of the Seychelles has also been sparked following recent East African successes in Mozambique and Tanzania.[58] Nearly 70 percent of the population of the Indian Ocean island nation of Mauritius is of Indian origin. The Indian oil company ONGC has sought oil and gas exploration rights in Mauritius, which has no known hydrocarbons.[59]

4. *West Rift Valley – Oil Discoveries in Uganda Bring Hope to Landlocked Nations.*

The West Rift Valley is a second branch of the East African Rift System, running geologically from Sudan, South Sudan, Uganda, eastern DRC, Rwanda, Burundi, western Tanzania, northeastern Zambia, Malawi, and western Mozambique. Only Uganda, the DRC, Rwanda, Burundi, and Malawi are addressed here.

Uganda is highly attractive to investors because the country's oil discovery rate is an extraordinary 90 percent. Estimates of Uganda's oil reserves have varied widely, with the major IOCs indicating 1 billion barrels and the government claiming 3.5 billion barrels. First, oil production has been repeatedly delayed because of disagreements between the government and IOCs; the current earliest start date is 2017, with full capacity unlikely to be reached before 2020. While there are some oil and gas fields in the narrow coastal strip of offshore DRC, the future of oil exploration for this troubled giant, Africa's second largest nation, lies principally in the east, in the Albertine Graben, where oil has already been discovered in significant quantities across the border in Uganda. Unfortunately, the DRC's government has mishandled the allocation of exploration blocks in eastern DRC, awarding some blocks as many as three times to different companies.

Rwanda has no oil production, but does have methane gas reserves along its borders with the DRC and Tanzania. Exploration is ongoing in the East Kivu Graben, located beneath Lake Kivu, which is part of the southern extension of the Albertine Graben in Uganda. Burundi announced in September 2012 that hydrocarbons had been found in Lake Tanganyika and the Rusizi plain. Some experts view the lake's geology as analogous to the tertiary rift system at Lake Albert to the north between the DRC and Uganda.

Malawi currently has no oil or gas production, but is part of the East African rift system, which is a proven exploration province with prolific oil discoveries in Sudan, Uganda, and Kenya. Until a long-standing border dispute with Tanzania can be resolved, exploration efforts on the contested section of Lake Malawi may be delayed.

5. *Central African Rift System – Complex Basin Systems, Rocked by Recent Instability Except for Chad.*

The Central Africa Rift System includes a series of basins running from Cameroon through southern Chad, northern Central African Republic (CAR), and Sudan, intersecting basins such as the Jongolei and Melut Basins that also extend into South Sudan. Also addressed in this section are two other countries in inland Africa, Niger, and Mali, which do not fit our model.

Chad was Africa's 10th largest oil producer in 2010. It shares with its neighbors three sedimentary basins known for oil and gas potential: 1) the Koufra–Erdis Basin in the north shared with Libya; 2) the Chad Basin in the center-west shared with Niger, northern Nigeria, and Cameroon; and 3) the Sud Basin in the south shared with CAR and Sudan. Since 2003, Chad has been exporting oil via a 1,000-km pipeline through neighboring Cameroon to the port of Kribi, on the Gulf of Guinea. In collaboration with Sudanese partners, China National Petroleum Corporation (CNPC) restarted oil exploration in 2011 in northeastern CAR near the border with Chad. Despite prospects for oil discovery, petroleum development and production in the CAR will likely be held back by the country's political instability.

Niger negotiated an agreement with Chad in July 2012 to use Chad's pipeline to export some of its future production. Niger has three distinct areas of oil and gas potential: Agadem, bordering Chad to the east; Tenere, with similar oil-producing basins as found in Libya, Chad, and Sudan; and Kafra, a continuation of a rift basin in northern Niger extending up to the Algerian border. Mali, to the west of Niger, is largely unexplored for hydrocarbons. Around one-half of the country overlies sedimentary basin, but only five exploration wells have ever been drilled. Further exploration in northern Mali is unlikely until the country's political situation stabilizes in the aftermath of a recent coup and attempted takeover of the north by a terrorist and secessionist groups.

6. *Southern Africa — At the End of Three Major Geological Regions, Plus Shale, Coal-Bed Methane Potential.*

Southern Africa is at the end of the Atlantic Transform Margin in the west, the West Rift Valley in south-central Africa, and the East African Coastal Rift in the east. The complexities of this geological sub-region also include the Karoo Basin shale oil in its west and coal-bed methane in its northeast. South Africa's oil and gas reserves are currently small, but could increase dramatically in the coming years. Offshore exploration in South Africa is intensifying, spurred by: 1) massive gas discoveries to the east in Mozambique in 2010, 2) modest discoveries in the Bredasdorp Basin to the south, and 3) encouraging results in the Orange Basin in the east, south of Namibia. Onshore, large IOCs are betting that the geological strata running southward from Uganda and the Great Lakes extend to South African territory.

Neither Botswana nor Zimbabwe has any proven oil or gas reserves, but both have potential to become producers of coal-bed methane gas, the latter only if its government creates a more hospitable environment for foreign investment. Zambia has no proven oil or gas reserves, but with oil discovered in Uganda and coal-bed methane in Botswana, Zambia is hopeful that commercially viable hydrocarbon deposits will be found. While Lesotho is in the Karoo Basin, geological analysis suggests that it is unlikely to be the site of commercially viable shale gas. Swaziland also has no proven hydrocarbons.

III. AFRICAN OIL AND GAS PRODUCTION AND RESERVES

Major African Producers in a Larger Global Context.

Africa is already a huge energy exporter, and its 570 million TOE exports represent 43 percent by dollar value of the continent's total exports. As shown in Table 1, five countries—in order of decreasing output—accounted for 82.3 percent of the continent's oil production in 2010: Nigeria, Angola, Libya, Algeria, and Egypt. Other smaller oil-producing countries are Cameroon, Chad, Congo, Cote d'Ivoire, the DRC, Equatorial Guinea, Gabon, Ghana, Sudan/South Sudan, and Tunisia, with a limited number of other countries such as Senegal and South Africa with daily production that is negligible.[60] In 2009, the major African natural gas producers were, in order of decreasing importance, Algeria, Egypt, Nigeria, and Libya. These figures for natural gas are dated, however, and in the last 4 years, Nigeria has started to monetize significantly much more of its natural gas production, while

Angola started exports in 2013 at its first LNG train after repeated delays.[61]

Oil: 2010	(000 bpd)	Gas: 2009	(bcf/d)
Nigeria	2,065	Algeria	7.88
Angola	1,790	Egypt	6.07
Libya	1,550	Nigeria	2.25
Algeria	1,250	Libya	1.54
Egypt	740	Equatorial Guinea	0.62
Sudan	480	Mozambique	0.35
Congo	270	Tunisia	0.35
Equatorial Guinea	255	South Africa	0.18
Garbon	245	Cote d/Ivoire	0.15
Chad	100	Angola	0.07
Others	237	Others	0.13
Total	8,982	Total	19.58

Source: "Africa Oil and Gas: A Continent on the Move," Ernst & Young, 2011, EYG no. DW0100 WR, no. 1107-1271904, available from *www.ey.com/GL/en/Industries/Oil---Gas/Africa-oil-and-gas--a-continent-on-the-move---Regional-prospects-and-opportunities-in-the-oil-and-gas-sector.*

Table 1. Ten Largest African Producers of Oil and Natural Gas.

With 54 countries on the African continent, along with the disputed area of Western Sahara, there are various metaphors that could be used to represent the rise of new oil producer nations in Africa. One author referred to long-established oil and gas producers such as Algeria, Libya, and Nigeria as "old lions," newer producers such as Angola as a "young lion," and new, emerging producers such as Ghana as "new cubs." Another industry observer focused on regional agglomerations of producer nations to describe three new "clubs": (1) Emerging West Africa—Angola deepwater (pre-salt), Ghana deepwater, Mauritania deepwater, Ivory Coast deepwater, Gabon deepwater; (2) East Africa—Mozambique deepwater, Tanzania deepwater, Kenya onshore and deepwater; and, (3) Onshore Central Africa—Chad, Uganda, Niger, Sudan, and South Sudan.[62]

This observer noted that several other countries that are not part of these clubs yet, such as Liberia, Sierra Leone, Somalia, South Africa, and Morocco, have seen new, successful basin tests and/or discoveries, but that these remaining countries have proven reserves that are of insufficient scale, with cumulative Africa totaling 600 million BOE, or just 1 percent of the total for Africa.[63] (The classification of these newly emerging producer nations such as Sierra Leone, and indeed perhaps two dozen of the countries described previously, thus enters into definitional issues of different categories of "reserves," an issue addressed in Appendix 2.)

Africa Boosting Oil and Gas Reserves.

As shown in Table 2, Africa had proven oil reserves of 132.4 billion barrels at the end of 2011, equivalent to 41.2 years of current production. This was an increase of 154 percent over the 1980 figure of 53.4 billion.[64] During the period from 2000 to November 2012, cumulative discoveries in Africa totaled 60-70 billion BOE (2P) or almost 10 percent of the total global additions of over 800 billion BOE, of which 380 billion BOE and 420 billion BOE, respectively, were conventional and unconventional resources. During this same period, new discoveries in emerging West Africa were 19.2-22 billion BOE, in East Africa 12.5-20 billion BOE, and in onshore Central Africa 3.2-4 billion BOE. More than half of Africa's new discoveries were added since 2007, with a large part being gas. Oil explorers in Africa had strong years in 2010 and 2011, when they found the equivalent of five billion barrels of oil each year, with 2012 being an even stronger year, with volumes approaching 10 billion barrels.[65]

It is also notable that the sources of these new reserves have become more geographically spread within Africa over time and that this trend toward increased geographical diversification should continue in the future. Recent discoveries in Africa have been made in all settings—onshore, below lakes, in shallow water, deepwater, in new countries and/or basins with little or no exploration history. Excluding new discoveries in North Africa, there were two major waves of exploration successes in the last 25 years. Nigeria was a leader in the 1990s in a series of deepwater discoveries, along with Angola, in proving 16 billion barrels of oil. A more recent wave, which is not reflected in Table 2, likely because of issues related to the definition of "proven reserves," can be dated to the 2000s and involves a new series of basins elsewhere in sub-Saharan Africa, including the Mauritania Basin in the country of the same name (2001), the Albertine Basin in Uganda, which straddles the DRC and the Kenyan part of the East African Rift (2006 for Uganda, 2012 for Kenya), the Tano Basin in Ghana (2007), the Rovuma Basin in Mozambique and Tanzania (2010 in both countries), and the Sierra Leone/Liberia Basin in both countries (2010 for Sierra Leone, 2012 for Liberia).[66]

Moreover, no unconventional reserves have been added to Table 2 despite the fact that the continent has substantial, proven heavy oil/bitumen in several countries, including Congo (Brazzaville), Nigeria, and Madagascar, and potential shale gas, most notably in South Africa, Algeria, Libya, Tunisia, and Ethiopia.[67]

At End 2011					
	Billion Tons	Billion Barrels	Percent Share of Total	Production 1,000 bpd	Reserves to Production Ratio
Algeria	1.5	12.2	0.7	1,729	19.3
Angola	1.8	13.5	0.8	1,746	21.2
Chad	0.2	1.5	0.1	114	36.1
Congo(Brazzaville)	0.3	1.9	0.1	295	18
Egypt	0.6	4.3	0.3	735	16
Equatorial Guinea	1.2	1.7	0.1	252	18.5
Gabon	0.5	3.7	0.2	245	41.2
Libya	6.1	47.1	2.9	479	*
Nigeria	5.0	37.2	2.3	2,457	41.5
Sudan and South Sudan	0.9	6.7	0.4	453	40.5
Tunisia	0.1	0.4	*	78	15
Other Africa	0.3	2.2	0.1	221	27
Africa Percent of World	7	8	7	11	

Source: *British Petroleum (BP) Statistical Review of World Energy,* June 2012.

Table 2. Oil-Proven Reserves by Leading African Country.

Although Africa's current oil reserves are only 8 percent of the world's total, the continent continues to play an outsized role "as a pillar of global energy security . . . providing a major source of diverse oil and gas supply."[68] While Africa has little excess capacity and thus cannot be considered a swing supplier per se, the continent's critical importance becomes most apparent when supply shocks take its production off the market, leading to immediate and large spikes in world oil prices. This occurred in 2008 after the Movement for the Emancipation of the Niger Delta (MEND) carried out attacks in Nigeria,[69] and in 2011 during civil strife in Libya.[70] Africa is also an important supplier region because it offers high quality oil, and a large number of producers not bound by OPEC quotas.[71] Additionally, Africa offers two other advantages that are highly relevant for the future: a region with relative freedom for IOCs to invest and operate,[72] and a high proportion of "new oil" projected to be discovered over the next 2 decades compared to other regions. With NOCs now managing roughly 80 percent of the world's oil supply—compared to the 1970s, when the IOCs controlled 85 percent of the world's oil reserves—the competition of Western, Chinese, and other emerging market oil companies in Africa is all the more intense; not surprisingly, with high oil prices and limited amounts of acreage available for exploration in other parts of the world, previously ignored African oil fields are attracting great interest from IOCs.[73] Moreover, much of the production in Africa is offshore. Although offshore oil is more expensive to develop, it is also less susceptible to the political instability that often affects onshore production on the continent[74] and less vulnerable to potential sabotage.[75] Offshore E&P also requires high technical capabilities—a strength of U.S. oil and gas companies. Not surprisingly, Africa has been a crucial element in the energy diversification strategies adopted in major consuming regions, notably Europe and China. Until the recent shale oil and gas boom, Africa had also been a major part of U.S. energy diversification, with U.S. oil imports from Africa for the first time surpassing those from the Middle East in 2007.[76]

According to 2012 estimates by BP and the U.S. Energy Information Administration (EIA) in Tables 3 and 4, Africa will be the world's fourth most important region for oil and gas production to 2035, after the Middle East and Europe/Eurasia, not far behind North America, and ahead of Latin America and East Asia/Australia. (Africa is likely to become an even more important region post-2035, but current projections are not usually made more than 25-30 years into the future.) BP predicts that from 2010 to 2030, there will be large increases in oil production in the Middle East, North America, and South/Central America, modest increases in Africa, and production declines in Europe/Eurasia, and Asia Pacific. In the case of Africa, BP predicts that African oil production will increase to 512.4 million tons of oil equivalent (TOE) in 2020, up from 478.2 million TOE, but then fall back to 497.6 million TOE in 2030. For African natural gas, the picture is unambiguously positive, with total production increasing from 188.1 million TOE to 257.2 million TOE in 2020 and 356.8 million TOE in 2030. Similarly, EIA predicted in 2012 that Africa oil production would rise to 11.45 million bpd in 2025, up from 10.75 million bpd in 2010, and then fall to 11.21 million bpd in 2035. The EIA's estimates for 2013 are much more optimistic about African oil, with production in 2035 now projected to climb to 12.09 million bpd in 2035, and to 12.64 million bpd in 2040 — the latter representing an 18.5 percent increase in daily production over the period 2010-2030.[77]

Million Tons of Oil Equivalent	2010	2015	2020	2025	2030
North America	648.2	700.1	736.9	757.7	788.3
South/Central Asia	350	395.5	426.5	450.2	477.7
Europe and Asia	853.3	836.7	823.6	800.8	790.5
Middle East	1,184.6	1,285	1,379.8	1,540.5	1,646.4
Africa	478.2	475.1	512.4	504.9	497.6
Asia Pacific	399.4	397.1	383.5	344	311.3
Total Oil Production	1,913.7	4,089.5	4,262.7	4,398.1	4,511.8
North America	750.4	795.7	852.7	875.6	906.2
South/Central America	145.1	175.1	191.8	233.9	270.7
Europe and Eurasia	938.8	1,016.5	1,047.9	1,092.5	1,161.8
Middle East	414.6	579	655.9	741.8	821.3
Africa	188.1	213.4	257.2	298.6	356.8
Asia Pacific	433.9	545.6	706.9	770.1	811.4
Total Natural Gas Production	2,880.9	3,325.3	3,712.4	4,012.5	4,328.2

Source: *BP Energy Outlook 2030*, January 2012.

Table 3. Global Oil and Gas Production (2010-2030).

	2010	2015	2025	2035
Middle East				
OPEC	23.43	25.46	29.77	33.94
Non-OPEC	1.58	1.43	1.18	0.97
Subtotal	25.01	26.89	30.95	34.91
North America				
United States	8.79	9.82	10.53	10.15
Canada	1.91	1.79	1.82	1.78
Mexico	2.98	2.65	1.58	1.68
Subtotal	13.68	14.267	13.93	13.61
Europe/Eurasia				
Russia	10.14	10.04	11.06	12.16
OECD Europe	4.36	3.7	3.15	2.83
Other Europe and Eurasia	3.22	3.67	4.37	4.54
Subtotal	17.72	17.41	18.58	19.53
Africa				
OPEC North Africa	3.89	3.62	3.37	3.37
OPEC West Africa	4.45	5.09	5.4	5.26
Non-OPEC Africa	2.41	2.4	2.68	2.68
Subtotal	10.75	11.11	11.45	11.21
Latin America				
OPEC South America	2.29	2.13	1.92	1.72
Brazil	2.19	2.72	3.87	4.45
Other Central/South America	2.01	2.29	2.47	2.65
Subtotal	6.49	7.14	8.26	8.82

Table 4. International Oil Supply by Region, 2010-2035 (in Millions of Barrels Per Day).

	2010	2015	2025	2035
East Asia/Australia				
China	4.27	4.29	4.79	4.7
Japan	0.13	0.14	0.15	0.16
Other Asia	3.77	3.79	3.38	3.0
Australia/New Zealand	0.62	0.55	0.54	0.53
Subtotal	8.79	8.77	8.86	8.39
Total OPEC	34.06	36.3	40.46	44.19
Total Non-OPEC	48.38	49.28	51.57	52.28
Grand Total	82.44	85.58	92.03	96.47

Note: Figures based on reference price for 2015, 2025, and 2035, which is between the "low oil price" and "high oil price" scenarios.

Source: *Annual Energy Outlook*, Washington, DC: U.S. Energy Information Administration, June 2012, p. 195, as modified to reflect regions.

Table 4. International Oil Supply by Region, 2010-2035 (in Millions of Barrels Per Day). (Cont.)

While these figures suggest only a modest increase in African oil and gas production until 2035, the author is encouraged about Africa's energy future because a deeper analysis of EIA's oil production estimates for Africa from 2010-2035 suggests that the continent's oil production could be far higher than current forecasts. Table 4 indicates that non-OPEC Africa (i.e., Africa's 50 countries other than the four OPEC members, Algeria, Angola, Libya, and Nigeria) oil production in 2010 was only 2.41 million bpd, or 22.4 percent of Africa's total, and will only climb to 2.68 million bpd in 2035, or 23.9 percent of the predicted African total. Given

Africa's large number of sedimentary basins outside of its four OPEC member-states, several of which are proven hydrocarbon provinces, and given the extremely low level of exploration in the remaining 50 African nations compared to these four "lions," it appears likely that EIA's estimates are highly conservative and mainly extrapolate from the present.

Africa's Competitive Landscape Becomes More Crowded—
Asian NOCs, Frontier Independents Join Traditional Supermajors.

What other evidence does the author have to corroborate his more optimistic views on future African oil and gas reserves? One leading indicator is that the competitive landscape for oil and gas exploration in Africa has become much more crowded over the last 10 years with the addition of Asian NOCs, smaller international independents, the creation of new NOCs in some African countries, and the rise of regional energy companies from Africa such as Nigeria's Oranto and South Africa's Sasol. For decades, the Western "supermajors" such as the U.S. firm Chevron, the United Kingdom's (UK) BP, and France's Total explored onshore acreage across the continent, taking advantage of tax and royalty concessions. In the 1990s, they started investing in deepwater, which at the time was considered anything deeper than 1,000 feet. They have been joined, however, by Russian NOC Gazprom and Asian NOCs from China (CNPC, the China National Offshore Oil Corporation [CNOOC], the China Petroleum and Chemical Corporation [Sinopec], and Sinopec's subsidiary Addax), India (ONGC), Indonesia (Pertamina), Korea (KNOC and its subsidiary

Dana Petroleum), Malaysia (Petronas), and Thailand (PTTEP), plus a limited number of Russian and Asian independent firms. Most of the growing and vibrant group of small-to-medium-sized independent firms is from Europe and the United States. These include Anglo-Irish Tullow Oil, France's Maurel et Prom and the U.S. firms, Anadarko and Kosmos. At least two dozen smaller independent E&P companies, some of which are privately held, have become active in Africa's so-called frontier countries, often where there had been little past exploration activity or where such activity dated to the 1960s or 1970s. These smaller players, some of which also come from the United States, are important because they have a narrowly focused exploration-led business model (in contrast to supermajors, which manage across the value chain from exploration to retailing), usually have much higher risk tolerance than supermajors, and are willing to develop smaller hydrocarbon deposits that may be uneconomic to bigger companies.[78] Examples of such firms include Canada's Africa Oil, and Australia's African Petroleum.

IV. GLOBALIZATION OF THE WORLD NATURAL GAS MARKET AND AFRICA'S ROLE IN IT

Up to this point, this monograph has focused mainly on **oil** production and reserves. However, **natural gas** will play a much bigger role in Africa's energy future than it does now, so it merits the following separate discussion.

At present, the **global** gas market is characterized by disparate **regional** markets. At present, the price of natural gas is much lower in the Atlantic Basin than it is in the Pacific Basin. At its lowest point in 2012, natural gas in the United States traded at around one-fifth of import prices in Europe and one-eighth of those in Japan.[79] Major suppliers, including Russia and Norway, have historically maintained high natural gas prices by limiting pipeline volumes into Europe, and, in the case of Qatar, by restricting spot LNG volumes to Asian buyers.

Recent events, particularly since 2010, however, are slowly creating a more competitive and interconnected global gas market. For example, more gas-related infrastructure development is enabling greater connectivity within and between regions, such as U.S. ports that can export LNG. Already, a limited amount of U.S. natural gas has been sold in Europe, and volumes could increase significantly if the U.S. Government grants more export licenses. The Panama Canal expansion, which will allow much larger PANAMAX LNG tankers, could also route U.S. LNG to Asian markets less expensively. Going forward, price differentials between regional gas markets will diminish as new supplier nations, many African, come on stream. LNG trade will become more competitive and contract terms will evolve, meaning that changes in one part of the world will be felt more quickly elsewhere.[80] Some argue that once the Panama Canal is expanded, LNG supertankers will bring the Atlantic Basin and the Pacific Basin into one global market—much as the advent of oil supertankers fostered a global market in oil.[81] In any event, West and Central African gas exports to Europe and Asia will help shrink the contract premiums significantly by 2020—providing a real boon to East Asian economies.[82] Additional gas supplies coming online in East Africa starting after 2018 will further compress the Pacific-Atlantic price differential.

The Present: Africa's Natural Gas Production and Reserves Focused on North and West Africa.

African gas **production** reached about 203 billion cubic meters (bcm) in 2011, with production by Algeria, Egypt, and Nigeria, collectively, accounting for more than 88 percent of the continent's total. Gas production in Africa since 2000 has been growing by about 4 percent per year.[83] Africa accounted for 6.4 percent of global natural gas production in 2010, but its share is expected to rise to 8.6 percent by 2035. As shown in Table 5, African gas production is projected to accelerate to 277 bcm in 2020 and 428 bcm in 2035.[84]

	1990	2010	2015	2020	2025	2030	2035
OECD	881	1,178	1,239	1,328	1,360	1,395	1,446
Americas	643	816	893	970	993	1,026	1,067
Canada	109	160	165	171	169	174	188
Mexico	26	50	45	51	57	66	75
United States	507	604	679	747	765	784	800
Europe	211	304	267	250	238	226	215
Asia Oceania	28	58	80	107	129	143	164
Non-OECD	1,178	2,106	2,377	2,616	2,908	3,215	3,509
East Europe/Eurasia	831	842	893	968	1,057	1,136	1,204
Asia	130	420	502	548	607	684	775
China	15	95	134	175	217	564	318
India	13	51	54	62	72	84	97
Middle East	92	472	565	609	660	722	809

Table 5. Global Natural Gas Production by Region.

	1990	2010	2015	2020	2025	2030	2035
Africa	64	209	221	277	346	402	428
Algeria	43	80	83	105	123	140	147
Libya	6	17	16	20	26	32	37
Nigeria	4	33	43	58	71	87	94
Latin America	60	163	195	213	238	271	292
World	2,059	3,284	3,616	3,943	4,268	4,610	4,955

Source: *World Energy Outlook 2012*, IEA, Table 4.5.

Table 5. Global Natural Gas Production by Region. (Cont.)

As shown in Table 6, the largest natural **gas producing** countries in Africa in 2010, in descending order, were Algeria, Egypt, Nigeria, and Libya. Africa's proven **reserves** of natural gas were estimated at 14 trillion cubic meters (tcm) as of January 1, 2012, or about 7.5 percent of the world's total. (Technically recoverable reserves of natural gas in Africa were estimated to be substantially higher, about 74 tcm, or almost 10 percent of the world's total.) The countries with the biggest reserves were, in descending order, Nigeria, Algeria, Egypt, and Libya, with these four countries accounting for more than 92 percent of the continent's total.[85]

Country	Reserves at January 1, 2012	2010 Demand	2010 Production
Algeria	4,452	28.5	83.7
Angola	307	0.7	0
Benin	1	0	0
Cameroon	134	*	*
Congo (Brazzaville)	90	0.9	0.9
Congo-Kinshasa	1	0	0
Egypt	2,162	45.6	60.6*
Equatorial Guinea	36	1.6	6.7
Ethiopia	25	0	0
Gabon	28	0.1	0.1
Ghana	22	0.1	0
Cote d'Ivoire	28	1.6	1.6
Libya	1,478	6.8	16.6*
Mauritania	28	0	0
Morocco	1	0.6	0.1
Mozambique	126	0.1	3.1
Namibia	62	0	0
Nigeria	5,053	4.9	28.7*
Rwanda	56	0	0
Senegal	0	*	*
Somalia	6	0	0
South Africa	0	4	1
Sudan	84	0	0
Tanzania	6	0.8	0.8
Tunisia	64	3.2	2
Total	14,264	99.5	206.6

*Note: 2011 production figures (bcm) for certain countries available as follows: Algeria (78.0), Egypt (61.3), Libya (4.1), Nigeria (39.9), and other Africa (19.4). Source: *BP Statistical Review of World Energy*, June 2012.

Source: "National Gas in Africa: The Frontiers of the Golden Age," Ernst & Young.

Table 6. African Natural Gas Production by Country (in Billions of Cubic Meters).

Historically, **North Africa** was the primary source of the continent's natural gas exports. Algeria has long been a major player in global gas markets and has been the second-largest gas supplier to Europe after Russia. Libya's natural gas is still relatively underdeveloped compared to oil, but the country is also an important supplier to Europe via the Greenstream pipeline to Italy.[86] Recent growth in Africa's natural gas supplies has come from **West Africa** through the huge associated gas developments that have accompanied an offshore oil boom led by Nigeria and Angola.[87] Additionally, West Africa, aided by the World Bank's Flaring Reduction Initiative, has become a focal point for capturing natural gas for export as LNG instead of being flared. While West Africa's growth in natural gas will continue as flaring is reduced and local gas infrastructure is developed, the big future for African gas lies in the massive offshore gas discoveries in **East Africa**, particularly in Mozambique and Tanzania.[88] Over time, the relative share of Africa's "big four" suppliers will decline as natural gas production in other countries such as Mozambique and Tanzania, comes online and as their reserves become reclassified as "proven."

East Africa and East Asia: A New "Golden Age" for Global Trade in Natural Gas?

Interest in the recent, large discoveries of natural gas in East Africa has been boosted by their convenient geographical position for LNG exports to Asian markets.[89] The Asian appetite for East African natural gas is also understandable because, as one analyst put it, "It's very important for Asian customers to secure gas free of the Strait of Hormuz risk."[90] Some analysts believe that final investment decisions to develop these East African gas reserves will not be reached, however, "without some long-term contracts attached, at least covering half or two-thirds of production."[91]

The ease of access to Asian markets and a break-even point that is substantially lower than rival Australia could help East Africa natural gas exports, particularly from Mozambique and Tanzania, according to consultant Wood Mackenzie.[92] The firm estimation as of August 2012 is that 100 trillion cubic feet (tcf) of gas has been discovered in Mozambique and Tanzania to date—enough for 16 LNG "trains" (as liquefaction facilities are called)—and that yet-to-be-found reserves were as much as 80 tcf in Mozambique and 15 tcf in Tanzania.[93] (The average discovery size in Tanzania has been much smaller, around 2 tcf, compared to Mozambique, which is over 7 tcf. Discoveries in Tanzania are also more spread out, so developing them will be more expensive than those in Mozambique because additional infrastructure will be required;[94] 100 tcf is enough to meet U.S. demand for 4 years.)[95]

Future Wild Card: Africa's Unconventional Gas Supplies.

Over the longer term, Africa's unconventional natural gas could also add substantially to the continent's potential supply.[96] In South Africa, the government recently lifted the moratorium on exploration and extraction of shale gas identified in three formations in the Karoo Basin.[97] Relatively new, "unconventional" supplies of natural gas in several African countries—including shale gas, tight gas, and coal-bed methane (also known as coal seam gas)—could add significant new supplies to the world's energy markets.[98] One analyst wrote that:

> So far, shale and other unconventional gas searches in Africa have been focused mostly in South Africa, but there is evidence of coalbed methane deposits in West Africa and some tight gas sands along West Africa's coast. Not without its environmental controversies, [Italian oil major] ENI's planned launch of a pilot project with an estimated 2.5 billion barrels of recoverable oil sands in the Republic of Congo, is the clearest sign yet that Africa north of South Africa and south of the Sahara may have more robust unconventional gas prospects.[99]

Tables 7 and 8 show, respectively, current and future planned LNG facilities in Africa, and current LNG export facilities under construction worldwide. The fact that there are far more planned African LNG facilities than current production facilities is consistent with rising expectations for Africa as a global supplier of natural gas.

Country	Project	Start*	Capacity	Operator
Existing/Operating				
Algeria	Arzew (3 trains)	1964	1.1	Sonatrach
	Skikda (4 trains)	1972	7.6	Sonatrach
	Bethioua (12 trains)	1978	16.5	Sonatrach
Egypt	Damietta (1 train)	2005	5.0	ENI
	ELNG (2 trains)	2005	7.2	BG Group
Libya	Marsa El Brega (2 trains)	1971	3.2	Sirte Oil
Nigeria	NLNG (6 trains)	1999	22.2	NNPC
Equatorial Guinea	Puna Eur (1 train)	2007	3.7	Marathon
Angola	Angola LNG (1 train)	2013	5.2	Chevron

Table 7. Current African LNG Production Capacity.

Country	Project	Start*	Capacity	Operator
Planned/Possible				
Algeria	Arzew GL 3Z	2013	4.7	Sonatrach
	Skikda LNG	2013	4.5	Sonatrach
Libya	Marsa El Brega T3	2013	2.6	Sirte Oil
Nigeria	Progress FLNG	2017	1.5	NNPC
Cameroon	Kribi LNG	2018	3.5	GDF Suez
Egypt	Damietta T2	2018	4.8	ENI
Equatorial Guinea	Punta Eur T2	2018	4.4	Marathon
Mozambique	Mozambique T2	2018	5.0	Anadarko
Nigeria	Brass LNG T1	2018	5.0	NNPC
Tanzania	Tanzania LNG T1	2018	6.6	BG Group
Mozambique	Mozambique T2	2019	5.0	Anadarko
Nigeria	Brass LNG T2	2019	5.0	NNPC
	NLNG T7	2019	5.0	NNPC
	NLNG T8	2020	8.5	NNPC
	OK LNG	2020	12.6	NNPC
Mozambique	Mamba	2020	10.0	ENI

*For existing projects, start date is date for first train; for planned/possible projects, start dates are nominal and subject to delay/cancellation.

Source: "Natural Gas in Africa: The Frontiers of the Golden Age," Ernst & Young, 2012, p. 14.

Table 7. Current African LNG Production Capacity. (Cont.)

Country	Project	Start-up	Mt/Y	Operator
Algeria	Skikda new train	end-2012	4.5	Sonatrach
	Gasi Touil LNG	2013	4.7	Sonatrach
Angola	Angola LNG	2013	5.2	Chevron
Australia	Gorgon LNG	2014-15	15.0	Chevron
	Queensland Curtis*	2014-15	8.5	BG
	Gladstone*	2015-16	7.8	Santos
	Australia Pacific **	2015	6.1	ConocoPhillips
	Wheatstone	2016-17	8.9	Chevron
	Prelude**	2017	3.6	Shell
	Ichthys	2017-18	8.4	Inpex
Indonesia	Donggi Senoro	2014	2.0	Mitsubishi
Papua New Guinea	PNG LNG	2014-15	6.6	ExxonMobil
Total			79.7	

Notes: *coal-bed methane based; **floating LNG project.

Source: *World Energy Outlook 2012,* International Energy Agency, Table 4.7.

Table 8. LNG Export Projects Under Construction Worldwide (July 2012).

V. IS AFRICA STILL RELEVANT TO U.S. ENERGY SECURITY?

It may seem provocative to ask the question, "Is Africa still relevant to U.S. energy security?" but it is a strategic question that must be posed. The short answer is that Africa **is** still of major importance to U.S. energy security, but that past projections of U.S. oil and gas imports from Africa are proving far too high.

Africa had provided up to 22 percent of U.S. oil imports in recent years, and the countries of the Gulf of Guinea alone had been projected by the U.S. National Intelligence Council to be the source of 25 percent of U.S. oil imports by 2015.[100] The new reality, however, is that the United States has been so successful in shale exploration and production in the last several years that it has been dramatically reducing its oil and gas imports from Africa, with 2012 U.S. oil imports from Africa being only 40 percent of their peak in 2007. Table 9 shows the long-term trend toward declining U.S. oil imports from Africa, which peaked in 2007.

Date	Algeria	Angola	Libya	Nigeria	Congo-Brazzaville	Congo-Kinshasa	Equatorial Guinea	Gabon	Ghana	Ivory Coast	Total
2003	40,992	132,349		303,617	9,755	698	21,452	47,670		1,368	557,901
2004	78,719	112,018	6,724	394,560	2,918	5,101	2,212	52,061		1,840	678,153
2005	83,359	166,404	15,934	393,038	9,216	688	24,734	46,515		7,238	747,126
2006	131,981	187,325	23,949	378,670	9,753		20,894	21,773		1,630	775,975
2007	161,770	181,813	30,794	395,554	23,123	473	20,070	22,897		166	836,660
2008	114,112	184,460	24,791	337,359	24,673		27,159	21,268		630	734,452
2009	102,559	163,604	22,354	283,091	23500	3,412	32,310	22,926		947	654,703
2010	119,579	139,736	15,608	358,924	25,694	3,225	18,225	17,022		3,130	701,143
2011	64,816	122,210	3,328	280,079	19,192	3,999	6,934	12,532	2,954	1,303	517,347
2012	43,891	80,945	20,358	148,353	10,710	137	15,100	15,462	150	1,394	336,500

Source: U.S. imports by country of origin, available from *www.eia.gov*.

Table 9. U.S. Imports of Oil by African Country of Origin (in Thousands of Barrels).

One reason for declining U.S. oil imports from Africa is that there has been a substitution in the U.S. import mix from African oil to oil produced in the Americas, including South America, Mexico, and Canada. Most importantly, however, there has also been a sharp absolute decline in U.S. oil imports from all source countries, primarily due to increased domestic production of nonconventional fuels, including shale oil and gas. The *Washington Post* analyzed this shift in oil imports in a January 2013 article, which stated that U.S. east coast refineries continued to import mainly African oil,[101] but that refineries in the Midwest were stopping West African imports:

> Over the past few years, U.S. refineries in the Midwest and elsewhere have spent billions upgrading their facilities in order to be able to process heavy, sour crude—the stuff that comes from the tar sands of Canada, or from Venezuela and Mexico. In order to recoup their investment, those refineries are likely to continue importing heavy oil from those three countries in the years ahead. By contrast, the light, sweet crude that's now pouring out of shale formations in North Dakota and Texas will mainly displace oil from Africa: Since July 2010, the U.S. has cut its Nigerian imports by half, from more than 1 million barrels a day, to 543,000 as of October 2012, the most recent data available through the EIA. Imports from Angola have dipped below 200,000 daily barrels, from an average of 513,000 in 2008. . . . By the second quarter of [2013], we will stop importing West African light, sweet crude into the Gulf.[102]

As shown in Table 10, there were only three African nations among the top 15 source countries for U.S. **oil** imports during the period January-November 2012: Nigeria (#5), Algeria (#10), and Angola (#11). These three countries cumulatively accounted for only 8 percent of U.S. imports. The United States imported about 40 percent less from Nigeria in 2012 than in 2011.[103] For the 9 years to 2011, the United States imported between 9 and 11 percent of its crude oil from Nigeria; in the first half of 2012, this figure fell to 5 percent.[104] U.S. **natural gas** imports have mainly been coming from Trinidad and Tobago, Qatar, and Algeria. Chevron has been forced to look at non-U.S. buyers, many in Europe, for its new natural gas production in Angola.

By Region	Thousand Barrels/Day	Percent of Total Imports	By Country	Thousand Barrels/Day
North America				
Canada	2,938	0.26	Canada	2,938
Mexico	1,032	0.09	Saudi Arabia	1,390
			Venezuela	939
Middle East			Mexico	1,032
Saudi Arabia	1,390	0.12	Nigeria	459
Iraq	475	0.04	Iraq	475
Kuwait	313	0.03	Colombia	436
Africa			Russia	473
Nigeria	459	0.04	Kuwait	313
Alergia	248	0.02	Alergia	248
Angola	243	0.02	Angola	243
Latin America			Equador	180
Venezuela	939	0.08	United Kingdom	156
Colombia	436	0.04	Brazil	237
Ecuador	180	0.02	Norway	79
Brazil	237	0.02		
Europe				
Russia	473	0.07		
United Kingdom	156	0.01		
Norway	79	0.01		

Source: EIA, February 1, 2013, p. 4, available from *www.eia.gov/petroleum/imports/companylevel/*.

Table 10. Total Imports of Petroleum into the United States by Top 15 Source Countries from January to November 2012.

U.S. imports of African oil and gas are likely to decline further in the coming years as well. The International Energy Agency (IEA) predicts that oil production in the United States will peak in 2020 at 11.1 million barrels a day, with U.S. oil imports falling within the next decade to four million bpd from the current rate of 10 million bpd. The U.S. Energy Information Administration (EIA) predicts that there will be only modest growth in demand for energy in the United States over the next 25 years, with an annual rate of growth in annual energy consumption of only 0.3 percent from 2010 to 2035. The United States is not projected to return to the levels of energy demand growth experienced in the 20 years prior to the 2008-09 recession because of more moderate projected economic growth and population growth, coupled with increasing levels of energy efficiency, according to ExxonMobil.[105] Around 2030, North America is forecast to become a net oil exporter[106] — an outcome that also means the United States will enjoy an order of magnitude greater energy security in 20 years than it does at present, and will have little dependence on Africa for its energy needs.[107]

Despite declining U.S. oil and gas imports, Africa is still of major importance to U.S. energy security — indeed to U.S. national security — for several reasons. First, although the United States will become significantly more energy independent in the years to come, it is still energy insecure for now, and Africa is thus still critical to a diverse supply of oil to the United States. Second, most of our Western allies, including Europe and Japan, will remain energy insecure for the foreseeable future, and will increasingly — not decreasingly — depend on oil and gas from Africa. Europe, for example, faced with declining oil and gas production in the North Sea and the unpredictability of Russia as an energy supplier, has increased its dependence on Africa, particularly on Algeria, Angola, Libya, and Nigeria for oil, and on Algeria and Libya for natural gas.[108] Third, the commercial stakes for U.S. investors in African oil and gas are already enormous and growing rapidly. U.S. strength, and ultimately national security, depends on a strong economic foundation that creates jobs and wealth through exports and successful foreign investment, including to Africa.

Table 11, a product of the author's country-by-country research found in Appendix 1, is a comprehensive but not exhaustive listing of current oil and gas exploration and production in Africa by U.S. companies. American firms have upstream activities in at least 22 countries on the continent, sometimes as the top first or second IOC in production. All but three of the 22 countries where U.S. firms are active in Africa are littoral or island nations. With the exception of the North African nations of Algeria, Libya, and Egypt, the large majority of active licenses held by U.S. firms involve offshore blocks, often deepwater. Outside of North Africa, sub-Saharan African countries with major U.S. oil and gas upstream investments include Nigeria, Angola, Equatorial Guinea, Chad, Ghana, and Mozambique. Other sub-Saharan countries with important U.S. upstream investments include Cameroon, Congo, Gabon, Kenya, Liberia, Madagascar, Morocco, Sierra Leone, South Africa, and Tanzania.

Country	Company	Details
Algeria*	Anadarko	Largest foreign producer with 500,000 bpd
* ConocoPhillips sold its Algerian operations to Indonesia's NOC Pertamina in December 2012	Hess	Gass E Agreb Redeve opment Project—nvest ng to boost product on n three ong-stand ng o fie ds
Angola	Chevron	Blocks 0 (Cabinda),2, 14. Soyo 5.2 Mmt/y LNG train. 434,000 operated bpd (36.4 percent share)
	ExxonMobil	394,300 operated bpd; Block 17 (20 percent share)
	Marathon	10 percent interest in Blocks 31 and 32.
	ConocoPhillips	30 percent operator interests in two deepwater offshore Blocks 36 and 37.
	Cobalt International	Operator rights to Angola's offshore blocks nearshore 9, and deepwater 20 and 21.
Cameroon	Kosmos	100 percent operated Ndian River Block and Fako Block
	Murphy Oil	Operator interests in three offshore blocks, 100 percent in Mer Profonde Nord and 50 percent in Mer Profonde Sud and 50 percent in Ndian River.
	Noble Energy	50 percent operator interests in the offshore YoYo mining concession and Tilapia exploration block.
	ExxonMobil/Chevron	Chad-Cameroon Pipeline one of the largest U.S. investments in sub-Saharan Africa, majority interest held by ExxonMobil and Chevron.
Chad	ExxonMobil (40 percent)/Chevron (25 percent)	Production peaked at 126,200 barrels per day in 2010, and averaged 115,000 bpd in 2011. Exporting oil via 1,000-km pipeline to Cameroon's Kribi port.
	ERHC	100 percent shares in three blocks in 2011,BDS 2008, Chari Ouest III, and Manga
Comoros	GX Technologies	Seismic studies of part of Comoros' territorial waters (probably only contractor, not investor)
Congo (Brazzaville)	Chevron	Deepwater Moho-Bilondo Field, 31.5 percent, non-operator share
	Murphy Oil	Operator of offshore blocks, Mer Profonde Nord (100 percent share) and Mer Profonde Sud (50 percent share)
	ExxonMobil	30 percent, non-operator share in the deepwater Mer Tres Profonde Sud
	Chevron	n jo nt Ango a-Congo L anz cross-border o fie d w th t e-back to BBLT p atform n Ango a s B ock 14

Table 11. Investment by U.S. Energy Companies in Upstream Assets in Africa.

Cote d'Ivoire	PanAtlantic (formerly Vanco	Non-operator in Block CI-401 (28.34 percent), block CI-524 (30 percent)
The Democratic Republic of the Congo (DRC)	EnerGulf	Operator of the Lotshi Block in the onshore salt Coastal Basin
	Chevron	17 72 percent non-operator share n 12 onshore and offshore fie ds south of Ango a s Cab nda and north of Angola's Soyo
Egypt	Apache	Largest producer of liquid hydrocarbons and natural gas in Western Desert, third largest in Egypt. In 2011, produced daily 217,000 bpd of crude oil and 865 mcf of natural gas (including joint venture partner-shares).
Equatorial Guinea	ExxonMobil	71.25 percent interest in Block B
	Hess	80.75 percent operator interest in the Ceiba Field
	Noble Energy	38 percent operator of b ock nc ud ng the 120 000 bpd Aseng o fie d and b ock O (45 percent operator) nc ud ng the A en gas-condensate fie d
	Marathon Oil	63 percent production sharing contract (PSC) for block A-12 offshore Bioko 60 percent interest in an LNG facility on Bioko Island and 40 percent
	PanAtlantic	80 percent operator Block K—Corisco Deep in Rio Muni, partial share of Block W, Block EG-02 offshore Bioko Island
	Brenham Oil & Gas (subsidiary of American International Industries)	15 percent participating interest for Block Y - offshore Rio Muni
	Murphy Oil	Operator Block W, offshore Rio Muni; in negotiations with government about acquiring interests in offshore Blocks J-14, J-15, K-14 and K-15
Ethiopia	Marathon	20 percent, non-operator share in license in Turkana Rift Basin
Gabon	VAALCO energy	50 percent operator in onshore Mutamba-Itoru exploration license
	Marathon	21.25 percent, non-operator in offshore Diaba license G4-223; 28.1 percent interests in exploration acreage in the Etame Marine
	Harvest Natural Resources	66.67 percent operator interest in Dussafu Marin Block
	Cobalt International	Diaba Block (share, operator status unclear)
	CAMAC energy	100 percent operator interests in two offshore blocks: A2 and A5

Table 11. Investment by U.S. Energy Companies

Country	Company	Details
Ghana	Kosmos Energy	30.87 percent operator of Deepwater West Cape Three Points block. 18 percent non-operator in Deepwater Tano Basin.
	Anadarko	30.87 percent non-operator of Deepwater West Cape Three Points block. 18 percent non-operator in Deepwater Tano Basin.
	Hess	47.22 percent operator interest in Deepwater Tano/Cape Three Points
	PanAtlantic	28.34 percent, non-operator interest in Cape Three Points Deepwater block
	Lushann International	Lushann-Eternit Energy has 55 percent share in Saltpond Field, whose current production is about 600 bpd
Guinea	Hyperdynamics	Holds one of the largest offshore exploration acreage positions in West Africa
Kenya	Marathon	50 percent interest in Block 9 and a 15 percent interest in Block 12A
	ERHC	Block 11A in northwestern Kenya
	Apache	50 percent operator in offshore Block L8
	Pancontinental	15 percent, non-operator Mbawa-1 exploration well
	Anadarko	50 percent operator of Blocks L5, L7, L11A, L11B and L12
	CAMAC Energy	100 percent operator of Blocks L1B, L16, L27 and L28
Liberia	Chevron	Acquired three offshore blocks, LB-11, LB-12, and LB-14, in which it currently has a 45 percent operator interest
	Anadarko	47.5 percent operator shares in Blocks LB-15, LB-16, and LB-17
	ExxonMobil	80 percent operator share offshore in Block LB-13
Libya	ConocoPhillips (16.33 percent), Hess (8.13 percent), and Marathon (16.33 percent)	In the Sirte Basin
	Hess	100 percent operator of Area 54 offshore in Sirte Basin
	Occidental Petroleum	Unknown share of Zueitina Field in Sirte Basin

Table 11. Investment by U.S. Energy Companies in Upstream Assets in Africa. (Cont.)

Madagascar	Madagascar Oil	100 percent of Blocks 3104-3107; Steam Flood Pilot project at Tsimiroro; 40 percent non-operator of Block 3102 (Bemolanga)
	ExxonMobil	70 percent interest in the offshore Ampasindava and Ambilobe Blocks in the northwest, and a 50 percent interest in the west-central offshore Majunga Block
Mauritania	MAREX	10 percent, non-operator interest in Deepwater Belo Profond
	Kosmos	90 percent operator interest in three offshore blocks
Morocco	Chevron	75 percent of three offshore areas: Cap Rhir Deep, Cap Cantin Deep, and Cap Walidia Deep
	Kosmos	Operator shares in four offshore blocks, three in the Agadir Basin, and the fourth in the Aaoun Basin. 56.25 percent participating interest in Foum Assaka Block, 75 percent interest in Essaouira Block
	Plains Exploration and Production	52 percent operator interest in the Mazagan permit
Mozambique	Anadarko	Discovered natural gas at its offshore Windjammer well in the Rovuma Basin, followed by a total of 10 successful exploration/appraisal wells also in the offshore block, "Area 1." Lead operator (35.7 percent) of Rovuma Onshore on the Tanzanian land border adjacent to the Indian Ocean
Namibia	EnerGulf Resources	Exploring for oil and gas offshore
	Duma Energy, Frontier Resources, and HydroCarb	Have encountered promising signs of hydrocarbons in the Owambo (Etosha) Basin. Frontier Resources has Blocks 1717 and 1817
Nigeria* * ConocoPhillips sold its Nigerian operations to Nigeria's Oando in December 2012	ExxonMobil	ExxonMobil is 2nd largest oil producer in Nigeria. Erha North offshore project.
	Chevron	Chevron is 3rd largest oil producer in Nigeria; 40 percent interest in 13 concessions with NNPC (NOC), and 18 to 100 percent shares in 10 deepwater blocks. Major shareholder, operator of Escravos Gasto-Liquids facility and West African Gas Pipeline (WASP).
	CAMAC (Exp. of several smaller U.S. independents in Nigeria)	CAMAC has an interest in a PSC for OML 120 with operator ENI of Italy.
Senegal	Fortesa	70 percent operator of Gadiaga Field; also has a license to explore the Tamna (Zone G) onshore block

Table 11. Investment by U.S. Energy Companies in Upstream Assets in Africa. (Cont.)

Sierra Leone	Anadarko	Recent discoveries, including 4 wells successfully drilled
	Kosmos	Block 4A
	Chevron/Noble Energy	Blocks 8a and 8b
	PanAtlantic	21 percent, non-operator in block in SL-5-11
South Africa	Anadarko	Exploration in Orange Basin
	ExxonMobil	75 percent operator interest in exploration rights in the Tugela south and three other offshore blocks
	Chevron	Seek unconventional exploration opportunities in Karoo Basin
Tanzania	ExxonMobil	35 percent, non-operator interest in Block 2
Zambia	Frontier Resources	Exploration Block 34

Source: Internet Research by author, primarily from January–April 2013. See country profiles for more details.

Table 11. Investment by U.S. Energy Companies in Upstream Assets in Africa. (Cont.)

54

Aside from exploration and production, commercial opportunities for future U.S. trade and investment in Africa's energy sector are extraordinary. IEA expects that almost $2.1 trillion will need to be invested in African oil and natural gas supply infrastructure over the period 2010-35—an average of more than $83 billion per year and more than will be invested in the Middle East, Latin America, or Asia over the same period.[109] One author predicted in 2009 that just the gross deepwater capital expenditure in West Africa alone between 2008 and 2015 would exceed that spent in Latin America, the Gulf of Mexico, the North Atlantic and Asia-Pacific.[110] Over the next five years alone, at least $230 billion will be required in Africa to develop LNG capacity and $22.6 billion to develop gas-to-power projects.[111] Much of this will be invested by U.S. operators including, for example, Chevron in Angola and Anadarko in Mozambique. U.S. oil services providers and industrial firms producing oil, gas, and power-sector equipment have traditionally been internationally competitive and, as such, Africa is a future bonanza that cannot be ignored.[112]

In the text above, the author asserts that, for now, Africa will remain important to U.S. "energy security," but that its importance will decline over the next 20 years due to greater oil and gas production in the United States and the rest of North America. What exactly do we mean here by "energy security"? Energy security has three components: reliability (or regular, uninterrupted access to energy), affordability, and environmental friendliness. At present, the United States enjoys a relatively high and improving level of energy security primarily related to **reliability** of supply.[1] The U.S. energy security is **high** not just because of booming shale oil and gas production, however, but also due to better mechanisms for dealing with supply challenges than in the past. These include the existence of a Strategic Petroleum Reserve started in 1975,[2] and the International Energy Agency created in 1994 to make possible timely collective responses to politically inspired disruptions in oil supplies.

U.S. energy security has also been enhanced by greater oil and gas imports from Canada, which is tied by pipeline to the U.S. energy economy.[3] The United States is also less vulnerable to disruptions in the global oil market in part because U.S. refiners have more refining capacity that is able to refine lower-quality, heavier crude, which means that they are not stuck with having to purchase higher-grade, lighter crude. (In contrast, the Chinese economy is structured around its crude oil production in Daqing, which produces relatively light, sweet, waxy crude. Chinese refineries are equipped to process this kind of crude and, since they have not converted many of their refineries to process heavier crude, they import significant amounts of light crudes into China.)[4]

Although the United States enjoys a relatively high level of energy security in 2013, plenty of risk remains. First, although U.S. dependence on oil imports is declining, our security vulnerabilities do not arise solely from the amount of oil we import. **Affordability** is also an issue.[5] Oil is a strategic commodity traded on a global market. Therefore, supply disruptions cause spikes in price, which then increase dramatically the U.S. import bill for oil, even though oil is still available to U.S. buyers. Second, for the United States, oil's strategic status does not stem from its use for electricity because only 1 percent of U.S electricity is generated from oil (other than in the strategic location of Hawaii, where the U.S. Pacific Fleet is headquartered, and where the large majority of electricity in that state is generated from imported fuel oil, about 70 percent of it from East Asia and Australia).[6] Instead, oil's strategic status stems from its virtual monopoly over the U.S. transportation fuel. In the United States, 98 percent of transportation energy is petroleum-based.[7]

How can the United States reduce its dependence on petroleum-based transportation energy? One way the United States has already reduced this dependence is to add more ethanol to our fuel mix. Another solution is to produce more vehicles that run on natural gas, thereby helping to monetize the price difference that exists today between oil and natural gas. This is already happening for U.S. truck fleets, especially those that constantly run the same routes because fueling stations are profitable to construct. Eighteen-wheel trucks running on natural gas save well over $150,000 in fuel costs over their 6-year expected life. Oil industry veteran T. Boone Pickens said in 2012 that converting U.S. truck fleets to natural gas could "knock out" 70 percent of U.S. imports from OPEC countries, and recalled that it took 6 years for the U.S. truck fleet to convert from gasoline to diesel in the 1970s.[8] Yet another option is to accelerate the move already started in the auto market toward hybrid gas-electric and all-electric vehicles. On a per-mile basis, electricity is much less expensive than oil.[9]

Fortunately, U.S. demand for oil is already flattening off — and U.S. energy security already increasing — in significant part because we are already improving efficiencies of our transportation fleet. Over time, this push toward greater fuel mileage will also be helped immensely by the Barack Obama administration's agreement in 2011 with 13 large automakers to increase fuel economy to 54.5 miles per gallon for cars and light trucks by 2025.[10] By around 2020, the United States will start to see the impact of the latest CAFE fuel-efficiency measures in transport. The result will be a continued fall in U.S. oil imports, to the extent that North America becomes a net oil exporter around 2030.[11]

This withdrawal of the United States as a major buyer in the global energy market (ex-North America) will also accelerate the switch in direction of international oil trade towards Asia, from Africa and the Middle East. The next two sections of this monograph will focus on Africa's importance to China as a source of energy, including how Beijing will likely build a blue-water navy whose core mission will include a presence in the Indian Ocean to protect oil and gas supply routes to its homeland.[12]

Endnotes - Text Box

1. The case can be made that unconventional oil and natural gas have also increased, on net, the environmental component of energy security. While hydraulic fracking, for example, carries risks of water contamination, the greater use of natural gas for power generation has displaced the use of heavily pollution coal.

2. See *en.wikipedia.org/wiki/Strategic Petroleum Reserve (United States)*.

3. Carolyn Pumphrey, ed., *The Energy and Security Nexus: A Strategic Dilemma*, Carlisle, PA: Strategic Studies Institute, U.S. Army War College, November 2012, p. 26.

4. *Ibid.*, pp. 28-29.

5. *Ibid.*, p. 2.

6. "Responses to Frequently Asked Questions Concerning Electricity Rates and Oil Prices," Honolulu, HI: State of Hawaii Public Utilities Commission, August 2, 2012; and "Fuel Use in Hawaii," available from *www.heco.com*.

7. Pumphrey, ed., *The Energy and Security Nexus*," p. 55.

8. "Pickens says converting truck fleets to natural gas will 'knock out' 70 percent of OPEC's imports," *DC Velocity*, June 28, 2012.

9. Pumphrey, ed., *The Energy and Security Nexus*," pp. 55, 58.

10. "Obama Administration Finalizes Historic 54.5 MPG Fuel Efficiency Standards," Washington, DC: The White House, August 28, 2012.

11. *World Energy Outlook 2012*, Paris: International Energy Agency (IEA), 2012, pp. 1-2.

12. *Ibid.*

Box 1. What is Energy Security and How to Promote it in the United States?

VI. AFRICA AND CHINA'S ENERGY SECURITY — AN INCREASINGLY IMPORTANT SOURCE OF DIVERSIFIED SUPPLY

Since the mid-1990s, Africa has played an increasingly strategic role in China's energy security. China's rapid economic development turned the country from a petroleum exporter to an importer by 1993 — a significant milestone in its development, and an event that spurred China to adopt a new foreign policy in 1995 emphasizing greater economic ties with Africa.[113] In recent years, the quality and availability of Africa's oil also became more important to Beijing as a way to offset declines in the North Sea and other light sweet crude producers that had previously met Chinese demand for crude and refined products.[114] As of 2009, Africa accounted for a significant share of the international equity production portfolio for China's "big three" oil companies: CNPC (40 percent), Sinopec (22.6 percent), and CNOOC (45 percent).[115] CNPC, the biggest of the three, also operates through its Great Wall Drilling subsidiary in eight African countries: Algeria, Chad, Kenya, Libya, Niger, Sudan, South Sudan, and Tunisia.

In 2009, China became the second largest net oil importer in the world behind the United States, with net total oil imports reaching 5.5 million bpd in 2011, versus 8.7 million bpd for the United States.[116] More than half of China's crude oil is imported, and about 33 percent of these imports came from Africa and 47 percent from the Middle East in 2010. Africa is a critical part of China's energy security diversification strategy, which seeks above all to reduce its dependence on Middle East oil. In December 2012, China overtook the United States as the world's largest net importer of oil, a development that the *Financial Times* predicted would:

> shake up the geopolitics of natural resources [since] the US has been the world's largest net importer of oil since the mid-1970s, shaping Washington's foreign policy towards energy-rich countries such as Saudi Arabia, Iraq and Venezuela.[117]

U.S. net oil imports dropped to 5.98 million bpd in December 2012, according to provisional figures from the EIA. In the same month, China's net oil imports surged to 6.12 million bpd, according to Chinese customs. By 2020, China will import about 65 percent of its crude oil. [118] An increasingly large share of China's imports will have to come from Africa.

Securing natural resources, most importantly oil and natural gas, is already Beijing's top national security priority in Africa.[119] While Africa's strategic importance for U.S. energy security is in relative decline due to shale oil and gas discoveries in America, the continent's importance continues to rise for China which, despite shale discoveries of its own, will increasingly depend on energy imports until well past 2040.[120] The IEA predicts, for example, that global energy demand will grow by one-third over the period to 2035, with China, India, and the Middle East accounting for 60 percent of the increase. Similarly, Table 12 predicts that energy demand in China will climb from 102 quadrillion British thermal units (QBTUs) in 2010 to 138 QBTUs in 2040 (and in India from 28 QBTUs in 2010 to 61 QBTUs in 2040, in the Middle East from 30 QBTUs in 2010 to 51 QBTUs).

Regions	Energy Demand (Quadrillion BTUs)			Percent Change			Share of Total		
	2010	2025	2040	2010-25	2025-40	2010-40	2010	2025	2040
World	525	633	692	21	9	32	100	100	100
OECD	227	234	224	3	-4	-2	43	37	32
Non-OECD	298	400	469	34	17	57	57	63	68
Africa	29	44	62	55	39	115	5	7	9
Asia Pacific	205	267	301	30	12	47	39	42	43
China	102	132	138	29	4	35	19	21	20
India	28	45	61	62	37	122	5	7	9
Europe	81	82	78	2	-4	-3	15	13	11
European Union	73	73	69	0	6	-6	14	12	10
Latin America	26	36	45	39	24	73	5	6	7
Middle East	30	42	51	41	21	71	6	7	7
North America	113	118	112	4	-4	-1	22	19	16
United States	94	96	90	2	-6	-5	18	15	13
Russia/Caspian	42	43	43	4	-1	3	8	7	6

Source: "The Outlook for Energy: A View to 2040," available from *www.ExxonMobil.com*, 2012, p. 49.

Table 12. World Energy Demand (2010-40).

As indicated in Table 13, China's top three sources of oil in the world in 2010 were Saudi Arabia, Angola, and Iran, while its top three sources of oil in Africa were Angola, Sudan, and Libya.

Saudi Arabia	893
Angola	788
Iran	426
Oman	317
Russia	284
Sudan	252
Iraq	225
Kuwait	197
Kazakhstan	184
Brazil	151
Libya	148
Other	922

Source: FACTS Global Energy, 2010.

Table 13. China's Oil Imports in 2010,
by Source Country
(in 1,000 Barrels Per Day).

Data from China's General Administration of Customs showed that Angola had become China's largest source of crude imports in April 2010, surpassing Saudi Arabia. As shown in Table 14, five of China's top seven trading partners in Africa in 2010 were oil exporters: Angola, Nigeria, Libya, Algeria, and Congo (Brazzaville).[121]

1.	South Africa	($25.7)	11.	Kenya	($1.7)
2.	Angola	($24.8)	12.	Tanzania	($1.5)
3.	Nigeria	($ 7.8)	13.	Ethiopia	($1.3)
4.	Egypt	($ 7.0)	14.	Mauritania	($1.2)
5.	Libya	($ 6.6)	15.	Gabon	($1.1)
6.	Algeria	($ 5.2)	16.	Tunisia	($1.1)
7.	Congo	($ 3.5)	17.	Equitorial Guinea	($1.1)
8.	Morocco	($ 2.9)	18.	Cameroon	($1.0)
9.	Zambia	($ 2.9)	19.	Chad	($0.8)
10.	Ghana	($ 2.1)	20.	Botswana	($0.4)

*Indicates country from which China purchased oil and minerals.

Table 14: China's Top Trade Partners in Africa in 2010 (in $ Billions).[122]

China's strategy of using Africa to move away from the Middle East/Strait of Hormuz risk through diversification has not been without its own risks. Libya had accounted for 3.1 percent of Chinese imports in 2010, but Chinese oil imports from Libya were disrupted by civil war and the North Atlantic Treaty Organization (NATO) intervention in March 2011.[123] The feud between Sudan and the recently independent South Sudan over oil pipeline fees also led to a multi-month cutoff in China's significant oil imports from those two countries. This being said, there is no doubt that, overall, China will enjoy much greater diversification of oil-source countries **within** Africa in the coming years due to ongoing exploration and production in multiple countries—both by Chinese and non-Chinese firms. This will enable China to be less vulnerable to disruptions within Africa from, e.g., an oil industry strike in Nigeria, and to disruptions elsewhere, e.g., a conflict in the Middle East tied to Iran.

China's Success in Marginal or Politically Sensitive Oil Fields.

Up to now, China's "oil diplomacy" in Africa has been most successful in niche markets such as Angola (due to corruption and to its $14 billion in infrastructure-for-oil deals),[124] and in other African countries with fields that have been ignored as too marginal or politically sensitive by major Western companies. Countries where China has invested in **marginal** fields include Nigeria, Chad, Congo, Niger, Gabon, Equatorial Guinea, and the DRC. In Niger, for example, China's CNCP took over a license in 2003 from an ExxonMobil/Petronas joint venture in Agadem that was not renewed.[125] Buttressing CNCP's investment in Niger, the China-Africa Development Fund (CADF) has underwritten a 2,000-km pipeline to export oil via another pipeline originating in Chad.[126] An example of a **politically sensitive** country is Sudan, where China's "no strings attached" foreign policy supported Beijing's oil diplomacy. In Sudan, both Chinese NOCs CNPC and Sinopec were able to take over projects from Western oil companies and ignore international sanctions. With the independence of South Sudan in July 2011, this new country became China's second most important source of oil in Africa.[127]

The Absolute and Relative Scale of China's Investment in Africa's Upstream Oil and Gas Industry.

Do oil-for-infrastructure in Angola and sanctions-busting in Sudan mean that the Chinese government's oil diplomacy is assisting Chinese energy companies to unfairly edge out traditional Western majors and smaller independents in Africa? In June 2008 congressional testimony, two U.S. Deputy Assistant Secretaries of State for Africa and East Asia said "no":

> There are often exaggerated charges that Chinese firms' activities or investment decisions are coordinated by the Chinese government as some sort of strategic gambit in the high-stakes game of global energy security. In reality, Chinese firms compete for profitable projects not only with more technically and politically savvy international firms, but also with each other.[128]

What do further analysis and the empirical data indicate? While Angola and Sudan may be exceptions, the author's research suggests that the U.S. officials were correct in their testimony. The Chinese Ministry of Commerce has commercial counselors in 48 of its Embassies in Africa, and they undoubtedly engaged in commercial diplomacy to lobby host governments in support of their NOCs — bolstered by relatively frequent, high-level visits by Chinese government or party officials. It is no coincidence that in March 2013 — a mere 4 months after assuming office — Chinese President Xi Jinping visited three African nations that are natural resource exporters: Congo (oil, with China as its biggest trading partner), Tanzania (an emerging natural gas exporter), and South Africa (a source of strategic metals and minerals).[129] This commercial diplomacy in no way implies, however, that the activities or investment decisions by Chinese NOCs in Africa are "coordinated by the Chinese government."

In Angola, a Beijing-sponsored $14.5 billion oil-for-infrastructure deal and corruption **may** have facilitated the acquisition by Chinese NOCs of non-operator shares in oil and gas blocks; in Sudan, Chinese and other Asian oil companies **did** purchase the assets of departing Western oil companies; and in Chad, diplomatic competition between Beijing and Taipei ironically **encouraged** oil sector investment by both the People's Republic of China (PRC) and Taiwan.[130] Despite these three, however, there are no other countries in Africa where Beijing's foreign policy priorities appear to have had a clear, determining influence on the activities or decisions of Chinese NOCs.[131]

In any event, regardless of whether the Chinese government's role has been more of advocacy or directive, Chinese oil companies **are** continuing to increase their investments in Africa rapidly, albeit from an initially modest base. Table 15, derived from the author's detailed country-by-country research presented in Appendix 1, is a comprehensive, but not exhaustive, listing of Chinese investment in upstream oil and gas assets in Africa. This table outlines 20-plus countries where Chinese firms reportedly have an interest in exploration and production. A perusal of this chart reveals that Chinese firms have a significant presence in three major oil and gas producing countries — Algeria, Angola, and Nigeria — and in four smaller producers — Cameroon, Chad, Gabon, and Sudan. Chinese oil firms also made major investments in Uganda in 2012 and Mozambique in 2013, only once as an operator (of one project in Uganda). The presence of Chinese oil companies in the remaining 11 African countries varies from as little as an office and rumored farm-in to new projects to, at best, exploration and limited production in others. If one compares the narratives in each country synopsis that follows and examines the maps of explorations blocks on the continent, the same picture of Chinese oil companies remains: growing significantly, but still dwarfed by the investments by Western majors and independents. (At the same time, and as becomes apparent in reading through the country-by-country profiles, it is notable that Chinese firms in Africa **do** have a relatively greater number of oil exploration and production licenses than firms from other, non-African developing countries, including the other BRIC countries of Brazil, Russia, and India.)

Country	Chinese Companies	Comments
Algeria	CNOOC	CNOOC, along with partner Sonatrach and Thailand's PTTEP, successfully discovered crude oil in September 2012 in the 4th and the 5th exploration wells in the Hassi Bir Rekaiz project.
	CNPC	Acquired the exploration licenses for Blocks 102a /112 in the Chelif Basin in northern Algeria (75 percent operator; 25 percent Sonatrach), Block 350 the northwestern Oued Mya Basin in the north of the Sahara Desert (75 percent operator; 25 percent Sonatrach), and Block 438b, also in the Oued Mya Basin (100 percent)
Angola	CNOOC	Non-operator shares in several offshore blocks.
	Sinopec	Joint venture with China Sonangol known as Sonangol Sinopec International (SSI), which holds 50 percent participating interest in Block 18, operated by the UK's BP. Equity production was 72,000 BOPD in 2010.
	China Sonangol	Besides JV with Sinopec, most important assets are 27.5 percent share in Block 17/06 and 20 percent share in Block 32. Also has non-operator interests in six other oil blocks in Angola (3/05, 3/05A, 15/06, 18/06, 31, and onshore Cabinda Block North).
Benin	Sinopec (Addax subsidiary)	50 percent interest in Block 1.
Cameroon	Sinopec (Addax subsidiary)	37.5 percent non-operator share n most prom nent new fie d, D sson . Struck hydrocarbons n October 2012 estimated at 20 million barrels of oil and 200 bcf of natural gas in the Padouk-1X Permit in Rio del Rey Basin. Ngosso license area is in Rio del Rey Basin (60 percent operator). Iroko license area is located approximately 30 km offshore (100 percent operator). 40 percent operator nterest n Loke e s Mokoko Abana F e d. 24.5 to 32.25 percent shares n e ght fie ds n Lokele (2) and Rio del Ray (6) operated by Perenco.
	Sinopec	Bought 80 percent stake in Shell Oil's Cameroon subsidiary in 2011; remaining 20 percent owned by NOC SNH. Purchase gave firm stakes n 12 offshore b ocks n Cameroon.
	Yang Chang Logone Development Holding	Acquired two blocks, Makary near Lake Chad, and Zina, about 20 km north of the Waza Park, in onshore northeast.

Table 15. Exploration and Production Activities of Chinese Energy Firms in Africa.

Central African Republic	CNPC	In collaboration with Sudanese partners, restarted oil exploration in 2011 in northeastern CAR. Ident fied four prom s ng sub-bas ns: Bagara (between CAR, Sudan, and South Sudan), Doseo (between CAR and Chad), Salamat (CAR), and Doba/Bangor (Chad).
Chad	CNPC	Bought shares in Block H, in whose Bongor Basin the company subsequently made several commercial discoveries starting in 2006. In 2007, signed agreement with Chadian Ministry of Petro eum for the estab shment of a 60-40 jo nt venture, 20,000-bpd refinery n Ndjamena, wh ch started product on n 2011. Construct on of 311-km p pe ne from fie ds to new refinery completed in 2011.
Congo	CNOOC	Hired a ship to drill in its offshore block in 2013 1H.
Equatorial Guinea	CNOOC	Block S.
	CNPC	70 percent operator interest in Block M in the Rio Muni Basin through Santa Isabel Petroleum subsidiary.
	Xuan Energy (Hong Kong)	Granted license in December 2012 as operator of newly created Block Y, offshore Rio Muni, with a 15 percent interest for Brenham Oil & Gas, a subsidiary of American International Industries.
Eritrea	Defba Oil Share Company, reportedly a joint venture between undisclosed Chinese company and Saudi-Bahraini Energy Alliance Company	Signed oil exploration agreements in 2008 for the two most northerly offshore blocks on the maritime border with Sudan—Bahri and Defnin.
Ethiopia	Sinopec	In 2007, the Ogaden rebel movement ONLF destroyed oil exploration facility outside Abole operated by Sinopec, killing 65 Ethiopians and nine Chinese.
	Petro-Trans (Hong Kong)	Malaysia's Petronas sold out its shares to China's Hong Kong-based PetroTrans in 2010, whose facilities were also threatened with attack from an Ogaden-based rebel movement in August 2011. Eth op an government cance ed five PSAs for 10 b ocks n Ju y 2012, c a m ng PetroTrans had fa ed to ra se prom sed financ ng for new nvestments.

Table 15. Exploration and Production Activities of Chinese Energy Firms in Africa. (Cont.)

Gabon	CNOOC	25 percent share of offshore Blocks BC9 and BCD10.
	Sinopec	Continued work on "several exploration wells" in Block G4-188.
	Sinopec (Addax subsidiary)	S nopec s Addax subs d ary had five product on shar ng agreements, two as onshore operator of Maghena and Panthere NZE, w th five produc ng fie ds, two d scover es, and one under development.
Ghana	Sinopec	In July 2012, won $750 million contract to develop Ghana's National Gas Processing Project, which involves transporting and marketing associated gas from the Jubilee Field. The Chinese government offered financ ng and agreed to accept repayment n the form of a ong-term o supply contract.
Guinea	China Sonangol	Offic a s n ear er Trans t ona government had awarded offshore acreage. However, democratically elected government of President Alpha Conde said he wiill review oil agreements made by previous regimes. China Sonangol's website does not list any oil and gas operations in Gu nea, but does st a Conakry office.
Kenya	CNOOC	In 2010, plugged the Bhogal-1 exploration well on Block 9 in north-central Kenya after inconclusive results, returned its two licenses, and left the country.
Liberia	CNPC (Hong Kong-listed arm, PetroChina)	Australia's African Petroleum made new offshore oil discovery in February 2013 when drilled Bee Eater-1 we . The company s fina z ng negot at ons w th PetroCh na for an nvestment of up to 20 percent in offshore Block LB-09.
Libya	CNPC	Partial interest with Libya's NOC in offshore Block 17-4 in the Pelagian Basin off the northwestern coast of Libya. Greatwall Drilling Company, a subsidiary, is also present in Libya Subsidiary.
Mauritania	CNPC	In 2006, found hydrocarbons in offshore Block 20, for which it has 65 percent operator interest, with Australia's Baraka Petroleum holding 35 percent. Onshore, exploration continues in the Taoudeni Basin, in central-eastern Mauritania.

Table 15. Exploration and Production Activities of Chinese Energy Firms in Africa. (Cont.)

Mozambique	CNPC	Italy's ENI reached an agreement in March 2013 to sell CNPC a 20 percent interest in offshore Area 4.
Niger	CNPC	In Agadem, took over in 2003 100 percent share in Block Bilma and an 80 percent share in Block Tenere, along with Canadian oil company, TG World (20 percent).
Nigeria	CNOOC	45 percent of Akpo Field (OML 130), operated by Total. 38 percent of Stubb Creek (OPL 229).
	CNPC	4 blocks: 2 in Niger Delta, one onshore, the other offshore (OPL 298 and OPL 471), 2 in Chad Basin in Borno Province in northern Nigeria (OPL 721 and OPL 732, the latter of which is a risk exploration block.
	Sinopec	Operatorship of NPDC's OML 64 and 66. 20 percent stake in the Usan Field.
	Sinopec (Addax subsidiary)	Onshore, 100 percent operator of OML 124. Offshore, 100 percent operator of OML 123, 126, 137; 72/5 percent operator of OPL 291, non-operator of Okwok (12 percent share) and OPL 227 (40 percent share) Non-operator of Block 1 (42.37 percent share) in JDZ with Sao Tome and Principe
Sao Tome and Principe	Sinopec	Operator in Block 2.
	Sinopec (Addax subsidiary)	Operates Blocks 3 and 4 in JDZ with Nigeria
	Sinoangol (not China Sonangol)	Applying for exploration rights in Block 2.
Sudan/South Sudan	CNCP	Shares in three consortia whose minority shareholder include Sudan NOC Sudapet and/or South Sudan NOC Nilepet.
	Sinopec	Sinopec (6 percent) and Egypt/Kuwait's Tri-Ocean Energy (5 percent) are minority shareholders in one of the three consortia, Dar Petroleum (Petrodar), which is based in South Sudan.
Uganda	CNOOC	In February 2012, Tullow farmed out part of interests to CNOOC and France's Total, with each taking a one-third share in Blocks EA-1, EA-2, and the former EA-3, now known as the Kanywataba and Kngfisher censes, w th each as operator of one of these three b ocks.

Table 15. Exploration and Production Activities of Chinese Energy Firms in Africa. (Cont.)

Where does the disconnect between popular misperceptions in the mainstream press of large-scale Chinese investments of African equity oil come from? One reason may be an erroneous extrapolation from overall foreign trade and investment trends between China and Africa, which do, indeed, show an exponential growth in commercial ties over the last 30 years. While this burgeoning economic relationship between Beijing and African capitals **is** real, it is diversified and not just concentrated in the energy sector. And while the **absolute** size of Chinese investment in African oil and gas assets is both growing significantly and likely part of a longer-term trend, one recent figure indicated that Chinese oil companies accounted for just 8 percent of the combined commercial value of all international oil company investments in Africa.[132] Why is this figure so low? One reason is because Western oil companies had a head start of several decades, making their cumulative investment much larger than that of China. Another reason is that these Western firms are not standing still, but instead investing huge amounts of capital into Africa's promising oil and gas sector, outweighing recent investments by Chinese NOCs.

At the same time, while the **relative** importance of Chinese NOCs in Africa may not be rising rapidly, the **absolute** amount of their investments certainly is. Chinese oil and gas multinationals have been actively investing in African assets via "farm-ins" of existing consortia, bids on licenses, and the outright acquisitions of smaller global independents.[133] CNOOC, along with France's Total, purchased in 2012 for $2.9 billion a **farm-in** of acreage in Uganda's Lake Albert Rift Basin controlled by Anglo-Irish Tullow Oil. This Tullow-Total-CNOOC partnership will spend between $10 and $12 billion by 2017 to develop around 3.5 billion barrels of oil resources in Uganda and export most of this to international markets. For its part, CNPC spent about $2.5 billion to buy a 20 percent stake in the Usan Field in Nigeria (Block OML 138) in November 2012[134] and $4.21 billion in March 2013 to farm-in to 20 percent of ENI of Italy's offshore Mozambique Area 4 gas fields.

Table 15 also provides multiple, recent examples of successful Chinese NOC bids on **licenses**. For example, in December 2012, the government of Equatorial Guinea granted Hong Kong-registered Xuan Energy a license as operator of newly created Block Y, offshore Rio Muni. Two examples of Chinese **acquisitions** of smaller independents with African ties are Sinopec's purchase in 2009 for $7.2 billion of Canada's Addax Petroleum (which had assets in several African countries) and CNOOC's $15 billion buy-out in early 2013 of Canada's Nexen (which had assets in Nigeria including in the Usan Field).[135] Sinopec is also rumored to be interested in buying some of the Gabon assets of France's Maurel et Prom, which account for 95 percent of the company's production, or in making an outright acquisition of the company.[136] Sinopec was also reported in February 2013 to be in talks with the UK's Afren about buying about $1 billion in oil assets in Nigeria.[137]

Is China's Investment in Africa's Oil and Gas Sectors Good or Bad?

Is rising Chinese investment in the Africa's oil and gas sectors good or bad? There certainly can be negatives from Chinese investment in Africa, such as questionable business practices that use bribery to undermine good governance, abusive labor practices that give preference to Chinese skilled and even unskilled labor, and lax compliance with international environmental norms and local laws. Another concern is the competitive advantage that Chinese companies have because they do not participate in the voluntary standards created by Western governments to foster improved governance, including environmental impacts and respect for human rights. Even for established oil producers like Nigeria, Angola, Equatorial Guinea, Gabon, and Congo (Brazzaville), there is a strong presumption that rent-seeking members of these governments have been corrupted by Chinese firms (and other Asian NOCs), which are not bound by OECD anti-corruption measures.[138]

Overall, however, the West should not be overly concerned about China's investment in Africa's energy sector. Why? For one, China's three major oil firms are actually leaders among their national brethren in other sectors in modernizing their business practices. Even China Sonangol, a past poster child for corrupt Chinese business practices in Africa, appears to be making considerable effort to repair the damage of past alleged actions through, e.g., hiring Western CEO Alain Fanaie, the former head of infrastructure and commodities at France's Credit Agricole,[139] and making its website, which now prominently touts the firm's work on corporate social responsibility, more transparent.[140] China's "big three" oil companies, while partially state-owned enterprises, are also traded on international exchanges, e.g., on the New York Stock Exchange as American Depositary Receipts (ADRs).[141] As such, they face certain disclosure requirements under U.S. securities laws. These "big three" oil companies are much more concerned about reputational risk than most other Chinese firms, and this positively influences their behavior in Africa.

The global energy market is also highly competitive, and China's three oil giants must compete with Western majors from a position of relative technological weakness. While China's companies are developing oil fields in Africa, for now they are not nearly as technically proficient as Western firms, especially in offshore exploration and production. For these reasons, Chinese oil companies have tended to be followers in Africa, not leaders. Angola is a case in point. For all of the media attention to the way China was able to win certain oil blocks and production "off-take" agreements in Angola, Chinese firms had interests in a mere 12 oil blocks—all only as non-operators. U.S. firms, by contrast, acted as operators of 11 oil blocks in Angola, and were also present as non-operators for five other oil blocks.[142]

Moreover, China's controversial use of nonmarket tools such as infrastructure-for-oil deals in Angola may be less unreasonable—and less repeatable in the future—than meets the eye. In Angola, China's EXIM Bank pledged massive, oil-backed financing of some $14.5 billion for Luanda's ambitious post-war reconstruction program, including over 100 projects in the areas of energy, water, health, education, telecommunications, fisheries, and public works. One Chinese academic asserted that Chinese government financing of infrastructure in Angola was not linked to deals by Chinese NOCs to win shares in oil blocks, but rather a means for Beijing to secure its loans to the government of Angola.[143] This explanation is at least partially correct. "Oil-for-infrastructure" is arguably merely a creative form of project finance that, for Beijing, also has the dual benefit of securing oil supplies. Moreover, this model may not be used again, at least within Angola. Some Angolans have complained that the infrastructure-for-oil deals with China have shackled their oil revenues to that of repaying Chinese loans.[144] Angola is also moving beyond the post-reconstruction phase of its history. Consistent with this, the Angolan government has looked to limit its exposure to further oil-backed loans with the Chinese, and the Chinese have not been particularly successful in Luanda in bidding for new oil blocks.

Another often raised concern is that China (and to a lesser extent, India) is somehow trying to negotiate long-term supply contracts in Africa to ensure that it can import oil and gas in the event of a crisis or supply disruption. In the case of oil, this concern is overblown; in the case of natural gas, it is outright misplaced. Other than the special case of post-conflict Angola, China has not cornered a significant amount of African **oil** at this time through other production off-take agreements, and is unlikely to emphasize this strategy in future deals on the continent.[145] Yes, there will be further infrastructure deals by China in the future, as there have been in the last 2 years in emerging oil and gas producers in Ghana and Tanzania, but these latest deals have not included oil off-take agreements. At the end of the day, much of the oil extracted by Chinese companies in Africa is not even exported back to China, but instead sold on the world market. Far from "cornering the market," Chinese energy companies (and those from other emerging countries) will continue to use benign farm-ins to existing energy projects led by Western majors and independents as their primary market-entry strategy. As working partners in these deals, they will continue to be in "learning mode," familiarizing themselves with the technology needed offshore for deepwater drilling and platform maintenance and skills needed onshore for regulatory compliance with African host governments. Over time, Chinese firms will increasingly act as lead operators in future deals, but still will not act to corner Africa's oil or gas markets.

Moreover, concerns about Chinese long-term supply contracts for **natural gas** are outright misplaced. In the case of natural gas, **any** long-term contracts supply contracts—assuming that they are arms-length, market-drive transactions not influenced by corruption—should be welcomed, not rejected, by African nations. Why? Any NOC and/or international consortia needs long-term purchase deals to reduce project risk and secure financing because major gas projects require huge capital investments in offshore and onshore infrastructure investments running into the tens of billions of dollars. This is precisely why, in March 2013, Italy's ENI farmed in China's CNPC into its Area 4 natural gas project in Mozambique, and why the U.S. firm Anadarko had earlier farmed in Japanese trading company Mitsui: to sew up long-term Asian customers and supply contracts that will make their projects feasible.

Does Africa Benefit from Chinese Energy Investment?

Again, from the African perspective, increased Chinese participation in Africa's energy sector should be seen as a big positive—**if** on fair, market-based terms. Chinese (and other Asian) players have bid up prices for acreage, brought new risk capital, invested in energy-related infrastructure, and found Chinese (and other Asian) customers for their oil and gas. It is good news from Africa's point of view that some Western oil industry participants have occasionally complained about China overbidding for acreage on the continent. Some Western observers have criticized these acquisitions of exploration blocks, asserting that some of them are economically unattractive or even technically unviable to develop.[146] A more logical explanation for investment by China's NOCs in Africa is not overbidding, however, but rather that they are behaving as rational economic actors. What may seem like marginal exploration blocks or existing production fields to a Western major with sophisticated risk models, precisely calculated "costs of capital," and disciplined capital budgeting processes may seem quite profitable to China NOCs, which enjoy lower personnel costs, lower cost of capital due to lending from state-owned banks, and thus lower "hurdle rates" for expected profitable projects.

The West, moreover, should also not be alarmed when Western oil firms sell their assets in Africa to Chinese (or other Asian) energy companies. Western firms, armed with the latest industry technology, are able to redeploy their capital on a global basis to exploit higher returns. At present, their most promising investments are as developers and operators of ultra-deepwater tracts in places like Angola and Ghana. This option is not currently available to Chinese NOCs, who still do not have the technical capacity to develop and operate ultra-deepwater offshore blocks, where much of the new oil and gas in Africa lies. Consequently, for now, Chinese NOCs have contented themselves to farming-in as non-operators in deepwater blocks and to bidding on onshore and shallower offshore blocks as operators. For example, in November 2011, China's Sinopec bought an 80 percent stake in Shell Oil's Cameroon subsidiary, with the 20 percent remainder owned by NOC SNH. This gave China stakes in 12 offshore blocks in Cameroon. For its part, Shell will redeploy the capital from this sale within its global business to locations with a higher expected rate of return or return the capital to its shareholders.

China and Rest of Asia: Biggest Winners from African Energy Expansion.

Aside from the **microeconomic** issue of the degree of involvement of Chinese NOCs in Africa's oil and gas sectors is the broader **macroeconomic** issue of how important Africa has been as the source continent for China's imported oil and gas. As noted previously, Africa was the source of 33 percent of China's **oil** imports in 2010, and this figure is likely to rise in the future. Even with the growth of its domestic **natural gas** production from shale and other nonconventional sources, China will still require over 130 bcm of natural gas imports by 2030. This is because while China's shale gas growth will accelerate after 2020, China's gas demand will increase from just over 150 bcm at present to more than 600 bcm in 2030. While much of China's imported natural gas may come from Russia, Turkmenistan,[147] Australia, and even North America, Chinese buyers will also seek diversified sources of natural gas from Africa, including from Gulf of Guinea countries such as Angola and Nigeria on the Atlantic side of the continent.

This monograph asserts that there has been a loose, inverse correlation between Africa's declining importance for U.S. energy security and its increasing importance for China. China and the rest of Asia have benefited **indirectly** but significantly from the increased supplies of shale oil and gas in the United States in the form of lower global prices, and will benefit **directly** the most from increased oil and gas supplies from Africa in the coming years. The changing energy security picture in North America has meant that new LNG projects originally targeting the U.S. market, such as that of Chevron in Angola, will be looking for Asian customers, including in China. IOC consortia leading natural gas projects in Mozambique and Tanzania have already started to accept Chinese and other Asian equity investment and are trying to lock in Chinese and other Asian customers. CNPC's $4.21 billion farm-in of 20 percent of ENI of Italy's offshore Mozambique Area 4 gas fields was the first such deal for China. Others will follow in the future.

One could argue that Japan will be the biggest beneficiary in Asia of these trends, because while its economy is only slightly smaller than that of China, its dependence on oil and gas imports is far higher. The author believes, however, that China will benefit even more over time than Japan from African energy because China's economy will likely continue to grow much faster than that of Japan and, as noted previously, China's dependence on energy imports will continue to increase both in absolute terms and relative to domestic supplies. Also, from a strategic point of view, the energy security of China's superpower rival in the 21st century, the United States, has increased dramatically because of the shale oil and gas boom, and will move toward relative energy independence as North American supplies increase. **When both China and the United States were big energy importers, they both felt vulnerable. In a few years, however, when the United States is more energy secure, China's own energy insecurity will be felt even more acutely by Beijing because its perceived main rival is more secure.** This is particularly so because the PRC has an autocratic government that is proud of its country's development, but also highly insecure and distrustful of the United States.[148] In this context, Africa will be not only a source of natural resources for China, but also an increasingly important source of **strategic** resources — oil and gas.

VII. AFRICAN ENERGY AND CHINA'S EMERGING TWO-OCEAN MILITARY STRATEGY

One academic team discussing African energy has asserted:

> In the strategic policy communities of both the United States and China, there has been a knee-jerk blurring between competition in commerce between U.S. and Chinese energy **firms** and the potential for strategic competition by one **country** to deny resources to the other.[149] (emphasis added)

Up to here, much of the author's research in this monograph has been presented largely in terms of the competition and cooperation between IOCs and NOCs — i.e., private and state-owned **companies** — from the United States, China, and other nations for exploration and production assets and licenses on the African continent. The strategic, country-level effort by China to secure its energy imports from Africa (and the Middle East) will now be examined.

An October 2012 Chatham House study referred to the oil trade with "Middle East exporters," but more accurately should have substituted this language with "**African and** Middle East exporters":

The axis of the oil market is shifting from the trade between the Middle East exporters and U.S. and European importers to one that links Asian developing markets to the Middle East, which no longer has sufficient oil to support these markets' growing needs.

With this shift, the study then posed the questions:

Who will carry responsibility for the physical security of Middle East oil export now that these mostly go to Asian markets rather than the US or Europe?[150]

The short answer is, in the short- and medium-term, the U.S. Navy will continue to protect the global commons around the Middle East (and Africa) for the benefit of the United States, its allies, and other non-ally countries that "free ride" on the security, such as China. However, that academic team also wrote that while:

At present . . . China has to depend on the United States and its allies to guarantee security of the main international shipping lanes and major 'choke points' from the Middle East through which so much of its imported oil comes. . . . Beijing is attempting to reduce this dependence on United States military power by increasing access to oil and gas imports from other regions, developing alternative routes and increasing its naval capacity.[151]

Alternative land routes that China is developing to import African and Middle East oil and gas have been collectively dubbed the "New Silk Road." This refers to China's efforts to help construct a modern grid of overland pipelines, roads, and railways from China to Central Asia and the Middle East so that a portion of its energy supplies can circumvent naval chokepoints and potential blockades or embargoes.[152] Other maritime-and-land routes contemplated or under construction include an oil pipeline from Gwadar, Pakistan, to Xinjiang in western China (not likely due to political instability),[153] an oil pipeline from the Bay of Bengal in Burma to China's Yunnan Province (gas pipeline opened May 2013, an oil pipeline opening in 2014, road/rail links opening later),[154] and a $20 billion or more canal or land bridge across the Isthmus of Kra in Thailand (feasible, but still in conceptual stage).

Over the longer term, a rising, wealthy, nationalistic and assertive China will also seek to increase its naval capabilities to secure its ever increasing energy imports from Africa and the Middle East and keep the Chinese Communist Party in power. Beijing's April 2013 Defense White Paper was quite clear on the linkage between its navy and energy security: "It is an essential national development strategy to . . . build China into a maritime power . . . Security issues are increasingly prominent, involving overseas energy and resources . . ."[155]

China has already been modernizing its navy and shifting its strategy from a "coastal defense" to a "far seas defense."[156] Military strategists have noted that China has developed a "two island chain" military doctrine in East Asia focused on its claims on Taiwan and the South China Sea. (See Map 1.) An ongoing evolution in this doctrine emphasizes not only the two island chains in the Pacific, but a "two oceans" strategy that includes the Indian Ocean.[157] The rising importance of the Indian Ocean in China's views of its national security derives from Beijing's increasing concern about: 1) transportation chokepoints for energy and other trade in the Straits of Hormuz, Mandeb, and Malacca; and, 2) the rise of India, with which it has had repeated border clashes. The "Malacca Dilemma"[158] refers to the fact that roughly 80-85 percent of China's oil imports traverse this key strait.[159] In its worst-case-scenario, Beijing is concerned about the risk of a U.S. naval blockade of China's energy imports should hostilities break out over Taiwan (or, even less likely, over the Senkaku/Diaoyutai Islands with Japan).[160] Furthermore, Washington's increasing energy security will make Beijing's own increasing energy insecurity be felt even more acutely, pushing the People's Liberation Army (PLA) to accelerate adoption of a "two ocean" military strategy that includes an enduring presence in the Indian Ocean as well as the Pacific.

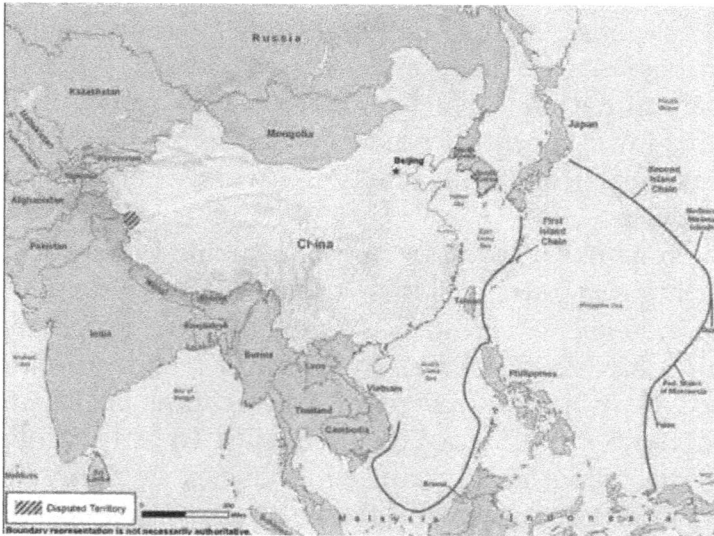

Source: T. X. Hammes, *Offshore Control: A Proposed Strategy for an Unlikely Conflict*, Institute for National Strategic Studies (INSS) Strategic Forum, Washington, DC, June 2012, p. 2.

Map 1. Two Island Chains in Military Doctrine of Peoples Republic of China.

What form will this "enduring presence" take? It is already clear that China is interested in building a regional support network of **refueling and resupply ports** in the Indian Ocean, sometimes referred to as a "string-of-pearls" strategy. This includes **South Asian** ports in littoral countries such as Pakistan (where Beijing has aided in constructing major port improvements at Gwadar), Sri Lanka (at Hambantota), Bangladesh (at Chittagong), and Burma, but also **African** ports, such as the Red Sea's Port Sudan in Sudan, Djibouti, at the Strait of Mandeb,[161] and the Seychelles, with which Beijing engaged in talks in December 2011 to establish a resupply facility, ostensibly to fight piracy.[162] There is also debate within China about whether to establish **military bases** overseas with infrastructure in the Indian Ocean and permanently stationed personnel. While some argue that China is unlikely to seek to establish a regional base network in the Indian Ocean, given the PRC's self-imposed policy against the "imperialist" basing of military forces in foreign countries, others (including the author) assert that this appears inevitable given China's energy needs. In support of the latter viewpoint, one academic recently argued that:

> While the Chinese government continues to proclaim its opposition to foreign bases, its practical actions and the words of its longest serving commander and the father of the modern [Chinese Navy] indicate otherwise.[163]

The author's own view is that China's recent participation in anti-piracy operations in Horn of Africa, in due time, will be recognized as a historical stepping stone toward a more active, permanent Chinese naval presence in the Indian Ocean. (See Map 2.) This presence may eventually include a naval base in close ally (and India arch-rival) Pakistan, but would mainly be comprised of a network of loose resupply arrangements in the Indian Ocean (including the Red Sea and Gulf of Aden) under the cover of supplying partner nations with either overall bilateral development assistance and/or more explicitly defense cooperation. In the future, these cooperative arrangements could be concluded with Sudan, Eritrea, Djibouti, the Seychelles, and Tanzania, for example. A January 2013 article in a Chinese state-run newspaper advised against building U.S.-style military bases, but advocated the establishment of a number of so-called "Overseas Strategic Support Bases." According to an English-language translation, the article indicated that:

> In the future, the Chinese Navy will establish its first batch of support bases in [the] Indian Ocean. . . . These bases can be divided into three levels: First ship fuel and material supply bases in peacetime such as the Port of Djibouti, Aden port of Yemen, and Salalah Port of Oman . . . the second is relatively fixed supply bases for warship berthing, fixed-wing reconnaissance aircraft, and the naval staff ashore rest, such as the ports of [the] Seychelles. China can build those bases by sign[ing] a short-term or medium-term agreement with [the] Seychelles; the third [are] fully-functional center[s] for replenishment, rest, and large warship weapons maintenance such as in Pakistan under medium- and long-term agreements.[164]

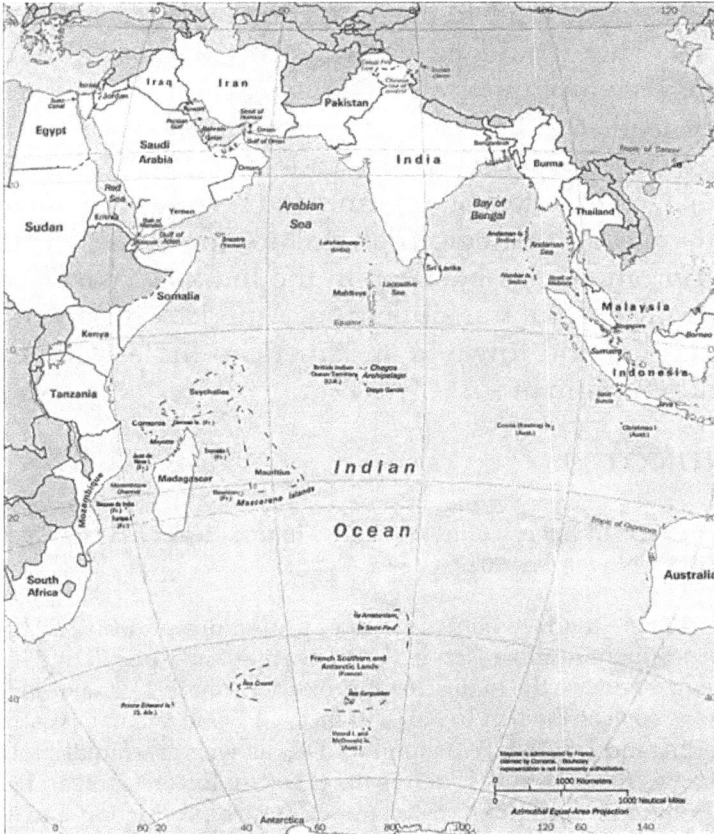

Source: *www.geographicguide.com.*

Map 2. The Indian Ocean, Including Strait of Malacca.

Whatever form it takes, over the long term, Beijing seems destined to develop a more muscular security cooperation policy vis-à-vis the Indian Ocean to ensure the security of its energy imports from the Middle East, East Africa, and the Gulf of Guinea in West Africa. While likely not welcomed by Washington or New Delhi, such a naval presence would not necessarily be inimical to U.S. or Indian interests, and could actually contribute to the protection of the global commons. This would particularly be the case **if**, some would say optimistically, China's peaceful rise really does lead it to be more of a responsible stakeholder in global prosperity. Some academics also argue that China, the United States, and India should work together in a "strategic triangle" to "preclude rivalry and promote collaboration in the Indian Ocean"[165] or seek "a peaceful transition from American unipolarity at sea . . . and towards an American-Indian-Chinese condominium of sorts."[166]

ENDNOTES

1. "World Energy Outlook 2012," International Energy Agency (IEA), November 2012, p. 135.

2. There tends to be two types of projections of future oil and gas production and reserves: conservative and speculative. On the one hand are the mainstream projections made by major institutions such as the IEA in Paris, France; OPEC in Vienna, Austria; the EIA and USGS in Washington, DC, as well as annual global estimates made by IOCs such as the U.S. firm ExxonMobil and the UK's BP. For sure, these estimates of future production and reserves are based on sophisticated and rigorous analysis, but they also tend to be conservative. On the other hand, there is a band of small, international exploration companies now hyperactive in Africa, mostly from Australia, Canada, the UK, and the United States. These small companies — the African analogues to the Texas wildcatters — are hungry for new discoveries and desperate for the huge amounts of capital needed for aeromagnetic, 2D, and 3D surveys, and for exploration, appraisal, and production wells. With these firms, there is a tendency to describe licenses or blocks as "highly prospective" and use "independent," third-party subject matter experts to corroborate these companies' optimistic projections. This monograph includes both these conservative and speculative projections.

3. "Continent," available from *en.wikipedia.org/wiki/Continent*.

4. "Geography of Africa," available from *en.wikipedia.org/wiki/ Geography of Africa.*

5. The seven island nations in Africa are Cape Verde, Sao Tome and Principe, Equatorial Guinea (which also has a mainland enclave), Madagascar, the Comoros, the Seychelles, and Mauritius.

6. Chakib Khelil, "A Framework for Investment in Africa's Energy Infrastructure," Oxford Energy Forum, Oxford, UK, November 2012, pp. 8-9.

7. "Is East Africa the Next Frontier for Oil?" *Time*, March 10, 2010.

8. Alex Vines, "Opportunities and Challenges in Africa's Changing Energy Landscape," Oxford Energy Forum, Oxford, UK, November 2012, p. 10.

9. Khelil, "A Framework for Investment in Africa's Energy Infrastructure," pp. 8-9.

10. "Africa Oil and Gas: A Continent on the Move," Ernst & Young, 2011, EYG no. DW0100, No. 1107-1271904, p. 3, available from *www.ey.com/GL/en/Industries/Oil---Gas/Africa-oil-and-gas--a-continent-on-the-move---Regional-prospects-and-opportunities-in-the-oil-and-gas-sector;* and Paul Collier, "The Plundered Planet: Why We Must—and How We Can—Manage Nature For Global Prosperity," New York: Oxford U.P., 2010.

11. See also Duncan Clarke, "Crude Continent," London, UK: Profile Books, 2008, for an interesting review of the history and politics of oil in Africa.

12. "Energy and Security in Africa: Meeting the Challenge in Petroleum Producing States," Abuja, Nigeria: Africa Center for Strategic Studies, March 6-11, 2005, p. 10.

13. "Campos Basin," available from *en.wikipedia.org/wiki/Campos Basin.*

14. "Lithosphere," available from *en.wikipedia.org/wiki/Lithosphere*.

15. Available from *en.wikipedia.org/wiki/Rift*.

16. Available from *en.wikipedia.org/wiki/Graben*.

17. "Pre-Salt Layer," available from *en.wikipedia.org/wiki/Presalt layer*.

18. "West Africa," available from *www.cobaltintl.com/assets/westafrica*.

19. Jacques Deverchere, "The North African Margin (Western Mediterranean Sea): Structure, Evolution, and Active Deformation," ESC 2010, September 6-10, 2010, p. 14.

20. J. D. Fairhead, "The West and Central African Rift Systems: Foreward," *Tectonophysics*, Vol. 213, 1992, pp. 139-140.

21. Kristjan Saemundsson, "The East African Rift System— An Overview," Short Course III, UNU-GTP, Lake Naivasha, Kenya, October 24-November 17, 2008.

22. Not discussed in this report are the French overseas department of Reunion, an island east of Madagascar; Mayotte, a French collectivity north of the Madagascar Channel; and the French islands of Iles Glorieuses, Brassas de india, which are also part of the Southern and Antarctic Lands and, like Juan de Nova, also found in the Madagascar Channel. See Map 2, available from *commons.wikimedia.org/wiki/File:France doutre-mer 2007 fr.png*, for more details.

23. James D. Fairhead, "The Mesozoic West and Central African Rift System: Qualitative Evaluation," *Search and Discovery*, Article #30077, 2009.

24. "World Energy Outlook 2012," p. 117.

25. *Ibid.*

26. Brian Swint, "Chariot Oil & Gas Plans More Namibia Wells After Two Dry Holes," New York: Bloomberg, September 21, 2012.

27. "Africa Oil and Gas," p. 5.

28. "Country Analysis Briefs—Equatorial Guinea," Washington, DC: U.S. Energy Information Administration, February 28, 2012.

29. "Gabon—Country Analysis Brief," Washington, DC: U.S. Energy Information Administration, January 17, 2012.

30. *Ibid.*

31. "Regions Petroliferes en Afrique" ("Petroleum Regions in Africa"), available from *fr.wikipedia.org.*

32. "Cameroon," available from *bowleven.com/Cam-Gab.*

33. "Cameroon—Expanding into Cameroon," available from *www.madisonpetro.com.*

34. "Africa Oil and Gas," p. 5.

35. "Benin: des gisements de petrole ont ete decouverts au large des cotes du Benin" ("Benin : Oil Deposits Discovered Along Coast of Benin").

36. See *africanpetroleum.com.au/our-projects/sierra-leone.*

37. *U.S. Geological Survey Minerals Yearbook—2011*, Advance Release, Washington, DC: U.S. Geological Survey, p. 35.5; Oil Exploration in Sierra Leone in 2011 . . . Maybe a New Frontier," June 14, 2011.

38. "Hyperdynamics Announces Updated Prospective Oil Resources Evaluation," February 14, 2013, available from *investors.hyperdynamics.com;* and "Tullow Agrees to Farm-in Guinea Concession," November 20, 2012.

39. David Brown, *The Challenge of Drug Trafficking to Democratic Governance and Human Security in West Africa*, Carlisle, PA: Strategic Studies Institute, U.S. Army War College, May 2013.

40. Available from *www.far.com.au/irm/content/senegal*.

41. Available from *www.mbendi.com/indy/oilg/af/gm/p0005*.

42. "Overview—Mauritania," available from *www.chariotoilandgas.com/operations/mauritania*.

43. "Historique de l'Exploitation" ("History of Exploration"), Morocco: Office National des Hydrocarbures et des Mines.

44. The Sahrawi Arab Democratic Republic (SADR) is a partially recognized state that claims sovereignty over the entire territory of Western Sahara, a former Spanish colony. SADR was proclaimed by the Polisario Front in 1976. The SADR government controls about 20-25 percent of the territory it claims, while Morocco controls the rest of the disputed territory. States that recognize the SADR's claims to sovereignty are mostly in Africa, though some non-African states do recognize it. It is currently not recognized by the United Nations (UN) as being a sovereign state. This information is available from *https://en.wikipedia.org/wiki/Sahrawi Arab Democratic Republic*.

45. "Oyster Oil Secures Djibouti Acreage," *Petroleum Africa*, September 20, 2011.

46. Angelia Sanders *et al.*, "Emerging Energy Resources in East Africa," Norfolk, VA: Civil-Military Fusion Center Mediterranean Basis Team, September 2012, p. 3.

47. "Country Profile: Kenya," Toronto, Canada: Cormark Securities, February 3, 2011, p. 15.

48. "Kenya to Gazette up to 9 New Oil and Gas Exploration Blocks," *Reuters*, February 12, 2013.

49. "Kenya," available from *www.apachecorp.com/Operations/Kenya*.

50. "Ophir Energy Reveals Jodari Field Discoveries," December 14, 2012; see *www.bg-group.com/Ourbusiness/WhereWeOperate/Pages/Tanzania*; "Papa-1 — Play Opening First Cretaceous Gas Discovery Outboard of the Rufiji Delta, Block 3, Tanzania," Ophir Energy PLC press release, August 2, 2012.

51. "Mozambique: Oil & Gas Exploration in 2011-2012," available from *mergersandacquisitionreviewcom.blogspot.com/2011/07/mozambique-oil-gas-exploration-in-2011.html*.

52. "Mozambique," available from *anadarko.com*.

53. "Oil, Gas and Security in Mozambique," available from *thinksecurityafrica.org/oilgas/oil-and-gas-in-Mozambique*.

54. "Mozambique Fact Sheet 2013," The Woodlands, TX: Anadarko Petroleum, 2013.

55. Debbie Stevenson, "Investors Make a Move on Madagascar's Oil Reserves," November 20, 2012.

56. "Madagascar," available from *www.nikoresources.com/operations/madagascar*.

57. "Oil, Gas and Security in Madagascar," available from *www.thinkafricasecurityafrica.org*.

58. "Offshore Seychelles a 'Flagship Project'," available from *www.whlenergy.com.au/irm/showStatic Category*.

59. "Mauritius: ONGC Considering Oil and Gas Exploration Rights," July 8, 2007, available from *energy-pedia general news*.

60. "Africa Country Profiles," available from *www.mebendi.com/indy/oilg/af/p0005.htm*.

61. EIA Short-Term Energy Outlook (STEO), as provided to author by USAFRICOM source on April 26, 2013.

62. Ivan Sandrea, "The Building Blocks of Africa's New O&G Industry," Oxford Energy Forum, Oxford, UK, November 2012, p. 3.

63. *Ibid.*

64. Rolake Akinkugbe, "Sub-Saharan Oil and Gas 2012: A Business Africa Guide," Lomé, Togo: Ecobank Capital, February 2, 2012.

65. Stanley Reed, "Finding Success on the Oil Frontier," *The New York Times*, July 3, 2012.

66. "Nigeria's Slack and the Rise of the Rest," *Africa Oil & Gas Report*, May 2012, pp. 4, 12.

67. Sandrea, "The Building Blocks of Africa's New O&G Industry," p. 3.

68. David L. Goldwyn, "Pursuing U.S. Energy Security Interests in Africa," prepublication draft, January 2009, p. 1.

69. Available from *en.wikipedia.org/wiki/Movement for the Emancipation of the Niger Delta.*

70. "Energy and Security in Africa: Meeting the Challenge in Petroleum Producing States," pp. 11, 16.

71. Despite periodic surges in resource nationalism in several African countries, Africa is a uniquely open market for foreign investment, particularly compared with Russia and Latin America, and has well over 50 percent of energy production coming from IOCs. See Goldwyn, "Pursuing U.S. Energy Security Interests in Africa," p. 7.

72. Bas Percival *et al.*, "Gambling in Sub-Saharan Africa — Energy Security through the Prism of Sino-African Relations," The Hague, The Netherlands: Netherlands Institute of International Relations, July 2009, p. v.

73. "High Prices Fuel Exploration Frenzy," *New African*, April 17, 2012.

74. Daniel Fikreyesus, "Oil and U.S. Foreign Policy Toward Africa," Ph.D. dissertation, Georgia State University, May 5, 2012, p. 9.

75. "Energy and Security in Africa: Meeting the Challenge in Petroleum Producing States," p. 24.

76. Of course, overall U.S. oil imports are declining but still important. As such, Africa remains important in diversifying U.S. energy import. See Fikreyesus, p. 1.

77. Calculations based on Table A21, "Annual Energy Outlook 2013," Washington, DC: U.S. EIA, p. 160, released May 2, 2013.

78. Monica Enfield, "Africa's Risk Outlook," Oxford Energy Forum, Oxford, UK, November 2012, p. 4.

79. *World Energy Outlook 2012*, Paris: International Energy Agency (IEA), 2012, p. 2.

80. *Ibid.*, p. 2.

81. Carolyn Pumphrey, ed., *The Energy and Security Nexus: A Strategic Dilemma*, Carlilsle, PA: Strategic Studies Institute, U.S. Army War College, November 2012, pp. 30, 32.

82. "Wood Mackenzie at Gastech 2012: Signposts Suggest a More Interconnected Global Gas Market in the Future," London, UK: Wood Mackenzie Media Centre, October 9, 2012.

83. Please note that, although Africa's natural gas production actually fell in 2011 compared to 2010 (to 203 bcm from 209 bcm), this was due primarily to supply disruptions in Libya due to the civil war in that country. With Libya excluded, natural gas production for the rest of the continent continued to increase. See also "Natural Gas in Africa—The Frontiers of the Golden Age," Ernst & Young, 2012, p. 6.

84. *World Energy Outlook 2012*, IEA, p. 139.

85. "Natural Gas in Africa—The Frontiers of the Golden Age," Ernst & Young, 2012, p. 6.

86. Elias Pungong, "Africa's Gas (R)evolution," Oxford Energy Forum, Oxford, UK, November 2012, p. 7.

87. "Natural Gas in Africa—The Frontiers of the Golden Age," Ernst & Young, 2012, p. 3.

88. Available from *www.ey.com/GL/en/Newsroom/News-releases/News Growing-role-for-Africa-in-the-Golden-Age-of-Gas.*

89. *World Energy Outlook 2012,* IEA, p. 134.

90. Guy Chazan, "East Africa: Embarrassment of Riches' Hard to Exploit," *The Financial Times,* October 8, 2012.

91. "Oil and Gas: East Africa's Race to Get Ready," *The Africa Report,* October 1, 2012.

92. "East Africa Seen As New Frontier for Gas: Report," AFP, October 30, 2012.

93. "East Africa Faces Technical and Commercial Challenges to Bring Its 100 tcf of Discovered Gas to Market," *www.woodmacresearch.com,* August 22, 2012.

94. Available from *www.lngworldnews.com/wood-mackenzie-100-tcf-of-gas-found-in-mozambique-and-tanzania-to-date/.*

95. Angelia Sanders *et al.,* "Emerging Energy Resources in East Africa," London, UK: Civil-Military Fusion Centre Mediterranean Basis Team, September 2012.

96. "Natural Gas in Africa—The Frontiers of the Golden Age," Ernst & Young, 2012, p. 3.

97. Pungong, p. 7.

98. "Natural Gas in Africa—The Frontiers of the Golden Age," Ernst & Young, 2012, p. 4.

99. Akinkugbe.

100. "African Oil: A Priority for U.S. National Security and African Development," undated, p. 6, available from *www.iasps.org/strategic/africawhitepaper.pdf*.

101. Another factor explaining the decline in U.S. oil imports from Africa is that two refineries on the U.S. east coast were idled in 2012, per EIA Short-Term Energy Outlook (STEO), as provided by USAFRICOM source on April 25, 2013.

102. Brad Plumer, "Where the U.S. Imports Its Oil From, in One Map," *The Washington Post*, January 19, 2013.

103. Joe Brock, "Analysis: Nigeria Losing Ground in Changing Oil World," *www.Reuters.com*, February 25, 2013.

104. "Nigeria," Country Background Note, U.S. Energy Information Administration, October 16, 2012.

105. See also *www.eia.gov/pressroom/releases/press374.cfm* for this analysis.

106. "Within the Next Decade U.S. Foreign Oil Imports Are Forecasted to Fall More Than 50%," December 6, 2012, available from *www.stockhouse.com*.

107. By 2035, imports into the United States are expected to be only 3.4 million bpd, and North America as a whole (including Canada and Mexico) may become a net exporting region. ExxonMobil predicts that energy use in OECD countries will remain essentially flat until 2040, while non-OECD energy demand will grow by close to 60 percent, making overall global energy demand about 30 percent higher in 2040 compared to 2010.

108. EU member-states purchase three-fifths of Algeria's LNG exports, while almost all of Libya's gas production is exported to Europe via the Greenstream pipeline under the Mediterranean to Italy. See Lord Aikins Aduei, "Global Energy Security and Africa's Rising Strategic Importance," January 6, 2012, available from *www.modernghana.com*; and the country profile on Libya.

109. "Africa Oil and Gas," p. 3.

110. Goldwyn, "Pursuing U.S. Energy Security Interests in Africa," p. 7.

111. Akinkugbe.

112. See also John P. Banks, "Key Sub-Saharan Energy Trends and their Importance for the U.S.," Brookings Africa Growth Initiative, March 13, pp. 8-13, for more analysis of the broader issue of the importance of energy for development. The author argues that "Failure [in Africa] to expand energy access, reduce energy imports, diversify energy sources and manage new-found oil and gas wealth for the benefit of society, especially the poor, directly impacts U.S. interests." The author lists these U.S. interests as humanitarian, national security, and economic.

113. David E. Brown, *Hidden Dragon, Crouching Lion: How China's Advance in Africa is Underestimated and Africa's Potential Underappreciated*, Carlisle, PA: Strategic Studies Institute, U.S. Army War College, September 2012, p. 29.

114. Monica Enfield, "Africa's Risk Outlook," Oxford Energy Forum, Oxford, UK, November 2012, p. 4.

115. Between 1994 and 1998, the Chinese government reorganized most state-owned oil and gas assets into two vertically integrated firms: CNPC and Sinopec. These two conglomerates operate a range of local subsidiaries and together dominate China's upstream and downstream oil markets. One additional state-owned oil firm that emerged over the last several years was the CNOOC, which is responsible for offshore oil exploration and production. See "China – Background," Country Analysis, Washington, DC: U.S. Energy Information Agency, September 4, 2012.

116. *Ibid.*

117. Javier Blas, "China Becomes World's Top Oil Importer," *The Financial Times*, March 4, 2013.

118. Shelly Zhao, "The Geopolitics of China-Africa Oil," China Briefing, April 13, 2011.

119. Coming in second to fourth in China's national interests in Africa are: China's commercial interests in promoting trade and investment on the continent, securing political support from African nations in the UN, and winning diplomatic recognition at the expense of Taiwan (the "One China Policy").

120. China's shale deposits are spread across four regions, including Sichuan and Shaanxi Provinces. Its shale deposits are more geologically complex than in the United States and found at depths of 3,000-5,000 meters compared with less than 3,000 meters for U.S. shale deposits. China's National Energy Administration forecasts shale gas output by 2020 of 60-100 bcm. See "Focus: Unconventional Oil & Gas—Europe, Africa, Asia Governments Assess Shale Development Policies," *Oil & Gas Journal*, available from *www.ogj.com/articles/print/vol-110/issue-07*.

121. Zhao.

122. "Sino-Africa Trade and Financial Cooperation: History and Trends," Bank of China, May 2011.

123. Zhao.

124. The financial, economic, and developmental aspects of China's oil-for-infrastructure deal were its primary attractions. Political considerations may also have been deciding factors, however. The Angolan government was able to rebuff demands from the International Monetary Fund for structural adjustments in 2004 because it secured this $14.5 billion loan from China. Old fashioned third-world and socialist solidarity between China and Angola's ruling party, the nominally Marxist Peoples Movement for the Liberation of Angola (MPLA), may have also been a factor in winning internal approval within the Chinese government and the acceptance of the Chinese by Luanda.

125. CNCP's acquired license includes a 100 percent share in Block Bilma and an 80 percent share in Block Tenere. The holder of the remaining 20 percent share was Canada's TG World.

126. Sanusha Naidu, "Africa Becoming Low Cost Manufacturing Hub for Chinese Investments," *Pambazuka News*, May 27, 2010.

127. Sudan and South Sudan also became a major diplomatic challenge for China. Most notably, a dispute between Sudan and South Sudan over oil export transport fees broke out, leading to a shutoff in oil exports in November 2011. Reflecting the importance of Sudan and South Sudan for Chinese energy imports, Beijing dispatched its special envoy for African Affairs to Khartoum and Juba in December 2011 to propose a solution to the dispute, dispatched a new envoy in early May 2012 in a renewed diplomatic effort, and joined other UN Security Council members that same month in unanimously supporting Resolution 2046, which called on Sudan and South Sudan to halt cross-border attacks and return to negotiations. (The Presidents of South Sudan and Sudan agreed in April 2013 to resume oil exports via the existing pipeline.) See "Sudan, South Sudan to Open Border, Resume Oil Exports," Associated Press, April 12, 2013.

128. As quoted in Christopher Alesi and Stephanie Hanson, "Expanding China-Africa Oil Ties," New York: Council on Foreign Relations, February 8, 2012.

129. "Xi Jinping Wraps Up Africa Trip in Congo," BBC News, March 29, 2013.

130. Chad, an inland nation where the investment risk-return facing Chinese NOCs in the early-2000s probably looked poor, may have been a case where the politics of winning diplomatic recognition from N'djamena (and pushing out Taiwan's NOC) may have induced Beijing to instruct its NOCs to make primarily policy-driven investments in Chad's prospective oil fields and in new pipelines and refinery capacity.

131. Another key point is that it is important to separate a discussion of China's NOCs in Africa from a discussion of China's government/Communist Party foreign policy on the continent. Of course, there is a close relationship between the NOCs and the Chinese state/party. This relationship is not only a formal ownership relationship under which the Chinese government, through its State-Owned Asset Supervision and Administration Commission, owns shares of the NOCs, which are hybrid organizations that are both state owned and partially owned by private Chinese and international stock investors. The key executives of the

"big three" Chinese oil companies are also all senior members of the Communist Party, and there is considerable rotation between the companies and the commissions and ministries that oversee them. Yet, these key executives are ultimately more appraised on their companies' profit performances than on their carrying out of Beijing's foreign policy objectives. Put differently, the political economy of most Chinese energy investments in Africa is far more about "economy" than "political."

132. Alex Vines, *Thirst for African Oil*, London, UK: London School of Economics, p. 20.

133. Besides China and aside from Middle East countries, other developing country players investing in Africa are from Russia (Gazprom, Lukoil, Rosneft), India (Oil India, ONGC), Malaysia (Petronas), and most recently, Indonesia (Pertamina and MedcoEnergi), Thailand (PTTEP), and Vietnam (PetroVietnam). See country profiles and "Natural Gas in Africa — The Frontiers of the Golden Age," Ernst & Young, 2012, p. 8.

134. Tara Patel, "Total Sells Nigeria Oil Field to Sinopec for $2.5 Billion," Bloomberg, November 19, 2012.

135. While the focus of this monograph is generally only on upstream exploration and production, China's important role in Africa's downstream oil and gas industry merits a footnote. Chinese firms have been building infrastructure across a wide range of oil-producing African countries — terminals, storage capacity, and, most critically, pipelines and refineries. In the case of refining capacity, the region's crude distillation unit (CDU) capacity is expected to increase by about 10 percent in the period leading up to 2015, with investment by Chinese investors dominating the proposed expansions. New Chinese-backed refineries are under construction in Chad and Niger, and have been proposed in Egypt, Sudan, and Nigeria. In addition, China's CNOOC is one of the partners (with Tullow and Total) in the proposed new refinery in Uganda, which will process the country's new crude oil production. China benefits not only from the profits of these contracts, but also the job creation and exports generating during construction of these projects. It also benefits to the extent that Chinese firms retain an ownership share in the refineries and pipelines. "H+K Strategies Advises CNOOC on $15.1 billion Acquisition of Nexen," *www.hkstrategies.com*, February 28, 2013.

136. "Maurel Prom Reveals Approach for Gabon Oil Assets, Report Names Suitor Sinopec," *www.energy-daily.com*, November 1, 2012.

137. Ziling We *et al.*, "Sinopec Said in Talks to Buy $1 Billion in Afren Assets," February 1, 2013, available from *bloomberg.com*.

138. Available from *csis.org/files/media/csis/pubs/csis africa review-energy prepbub draft.pdf*.

139. "China Sonangol Shows Its Hand," *www.Africa-Asia-Confidential.com*, December 2012.

140. "Corporate Social Responsibility," from China Sonangol. Some observers would contend that there remains a significant gap between its stated commitment to CSR and the realities on the ground.

141. CNPC, de facto, trades as an ADR in New York through its main subsidiary, PetroChina.

142. "Trends in U.S. and Chinese Economic Engagement," GAO-13-199, Washington, DC: U.S. General Accountability Office, February 2013, p. 43.

143. Anonymous source, conference on China in Africa, Washington, DC, February 14, 2013.

144. Matthew Plowright, "China & Africa: Law of the Land," September 6, 2011, available from *www.emergingmarkets.org*.

145. David Goldwyn, "Energy Security: The New Threats in Latin America and Africa," *Current History*, December 2006, p. 442.

146. Goldwyn, "Pursuing U.S. Energy Security Interests in Africa," p. 9.

147. China's major natural gas pipeline for imports is the Central Asian Gas Pipeline (CAGP), which spans 1,130 miles, has a capacity of 1.4 tcf/y and brings natural gas to China from Turkmenistan, Uzbekistan, and Kazakhstan.

148. The case can be made that the boom in shale gas in the United States, along with anticipated new supplies in Africa, has also strengthened China's negotiating hand with suppliers in its "near abroad," such as Russia. In 2006, CNPC officials signed a Memorandum of Understanding with Russia's Gazprom for two pipeline proposals, one from Russia's western Kovykta gas field to northwestern China with a pipeline capacity between 1 and 1.4 tcf/y by 2015. A second proposed route, called the Eastern pipeline, would connect Russia's Far East and Sakhalin Island to northeastern China and would have 1.1 to 1.4 tcf/y of capacity. The countries have yet to agree on a price for the gas, however. Even before this, Russia had been trying for more than 10 years to negotiate long-term gas supply agreements with China that try to match the high prices that it had extracted from Europe for long-term contracts that pre-date the North American oil shale boom and the new gas discoveries in East Africa. The Chinese have dug their heels in, insisting on much lower prices, and know that time is now on their side. In March 2013, long-term natural gas contracts were at the top of the agenda in talks between Russian President Vladimir Putin and newly installed Chinese President Xi Jinping. The author believes that Russia may ultimately reach deals with China (and Japan) on more attractive terms than offered in the past because of this greater competition among suppliers. See "China—Background," Country Analysis, on pipelines; and "Russia, China Inch Closer to Natural Gas Deal," *Ria-Novosti*, March 22, 2013, for details on the Putin-Xi meeting.

149. Angelica Austin *et al.*, "Energy Interests and Alliances: China, America, and Africa," Policy Paper 7/2008, Brussels, Belgium: EastWest Institute, August 2008, p. i.

150. John Mitchell *et al.*, "What Next for the Oil and Gas Industry?" London, UK: Chatham House, October 2012, p. 2.

151. *Ibid.*, p. 20.

152. Christina Lin, "China's Strategic Shift Toward the Region of the Four Seas: The Middle Kingdom Arrives in the Middle East," *Middle East Review of International Affairs*, Vol. 17, No. 1, Spring 2013, p. 33.

153. Lieutenant Colonel Daniel J. Kostecka, "A Bogus Asian Pearl," *U.S. Naval Institute Proceedings*, April 2011, pp. 50-51.

154. David Pilling, "Pipeline marks scramble for Myanmar," *Financial Times*, January 30, 2013.

155. "The Diversified Employment of China's Armed Forces," State Council, April 2013, available from *news.xinhuanet.com/english/china/2013-04/16/c 132312681.htm.*

156. *Ibid.*, p. 33, Note: The original Chinese, 远海 (yuanhai), has also been translated as "*distant seas.*"

157. Robert D. Kaplan, "China's Two-Ocean Strategy," Chap. II, Abraham Denmark and Nirav Patel, eds., *China's Arrival: A Strategic Framework for a Global Relationship*, Washington, DC: Center for a New American Security, September 2009.

158. *Ibid.*, p. 51, makes the following historical analogy with the United States that is useful for understanding the importance of the Malacca Strait to China:

> Just as the U.S. Navy moved a century ago to control the Caribbean Basin, so must the Chinese Navy move, if not to control, then to at least . . . become as dominant as the Americans in these seas: for the Malacca Strait can be thought of akin to the Panama Canal, an outlet to the wider world.

159. Lin, p. 32. Note: Lin has a figure of 80 percent, while Kaplan, p. 49, indicates it is as much as 85 percent.

160. Regarding 2), it is notable that India's naval forces in the Indian Ocean are becoming much more powerful and already include "state-of-the-art aircraft carriers, nuclear submarines, and other surface combatants." See Jonathan Holslag, "China's New Security Strategy for Africa," *Parameters*, Carlisle, PA: U.S. Army War College, Summer 2008, p. 26.

161. Daniel J. Kostecka, "Places and Bases—The Chinese Navy's Emerging Support Network in the Indian Ocean," *Naval War College Review*, Winter 2011, Vol. 64, No. 1, pp. 68-70.

162. Wilson Chau, "Seychelles Invites China to Set Up Naval Base," San Francisco, CA: New Pacific Institute, December 7, 2011.

163. Loro Horta, "China Turns to the Sea: Changes in the People's Liberation Army Navy Doctrine and Force Structure," *Comparative Strategy*, Vol. 31, No. 393, p. 400.

164. Jorge Benitez, "International Herald Leader," Chinese state-run newspaper under Xinhua News Agency, China Defense Mashup, as translated in "China Paper Urges PLA Navy to Build Overseas Military Bases," *www.acus.org*, January 19, 2013.

165. James R. Holmes *et al.*, "China and the United States in the Indian Ocean—An Emerging Strategic Triangle?" *Naval War College Review*, Summer 2008, Vol. 61, No. 3, p. 42.

166. Kaplan, p. 57.

APPENDIX I
AFRICAN OIL AND GAS EXPLORATION
AND PRODUCTION –
COUNTRY PROFILES LISTED ALPHABETICALLY

Algeria.

Depending on information source, in 2010, Organization of Petroleum Exporting Countries (OPEC) member Algeria was either the world's 8th or 10th largest producer of natural gas and 16th largest producer of crude oil. The country accounted for 2.5 percent and 2.0 percent of world natural gas and crude oil output, respectively. In 2010, Algeria was Africa's leading producer of natural gas and third or fourth-ranked producer of crude oil after Nigeria and Angola, and ahead or behind of Libya, depending on information source.[1] Algeria held 159 trillion cubic meters (tcm) of proven natural gas reserves, 2.4 percent of the world's total, the 10th largest reserves in the world, and second largest in Africa after Nigeria. Algeria had 12.2 billion barrels of proven crude oil reserves, or 0.9 percent of the world's total.[2] Reserves were the third largest in Africa behind Libya (47.1 billion barrels of oil [bbbl] and Nigeria (37.2 bbbl). In 2011, Algeria's crude oil exports were 750,000 barrels per day (bpd), of which 40.5 percent went to North America, mainly to the United States.[3]

Of Algeria's current hydrocarbon output, 75 percent comes from two fields: Hassi Messaoud and Hassi R'Mel. Algeria's NOC, Sonatrach, was responsible for 72 percent of hydrocarbon output; international oil companies (IOCs) working under Production Sharing Agreements (PSAs) were responsible for the remainder. These IOCs included U.S. firm Anadarko (the largest foreign producer, with 500,000 bpd),[4] a number of European IOCs, and other emerging market NOCs, including China's CNOOC,[5] CNPC,[6] Russia's Gazprom,[7] and Thailand's PTTEP.[8] The U.S. firm ConocoPhillips announced in December 2012 that it would sell its Algerian operations to Indonesia NOC Pertamina for $1.75 billion.[9]

Algeria's unfavorable 2005 hydrocarbon law and a 2006 presidential decree have dampened the enthusiasm among IOCs for investment in the country's oil and gas sector, contributing to a 20 percent drop in hydrocarbon production over the last 5 years.[10] Because of meager interest, the Algerian government only awarded three of 10 permits offered in a 2010 round, including an award to Spain's Repsol, and two of 10 in another round in 2011, with only France's Total a winner among international majors. In January 2013, Algeria's parliament approved amendments to the law that cancelled a windfall tax on foreign firms and replaced it by a "complementary tax on results." It also offered incentives to foreign firms to invest in unconventional resources.[11] The government announced that it plans a new round of bidding on permits in 2013, following the adoption of the new hydrocarbon amendments.[12] Some observers, however, remain pessimistic that these amendments will be able to turn around declining foreign investment and production because of regulatory instability and Algeria's low ranking—148 out of 183 nations—in the World Bank's "Doing Business" index.[13] A number of fields operated by Sonatrach are in steep decline, including Hassi Berine South (with Anadarko) and Rhourde Oulad Jemmah (with Australia's BHP Billiton).[14]

Another serious problem is that Algeria's domestic demand for oil and gas is consuming an increasing proportion of production. Domestic oil consumption rose from 26 percent of production in 2005 to 40 percent in 2010; domestic consumption of natural gas rose from 19 percent of production to 29 percent. A 1.5 percent annual increase in population, a rapid increase in the number of cars, and artificially low domestic prices were all factors behind this increased domestic demand.[15] Without increased investment, these trends led a Chatham House study to warn that Algeria would have no oil available to export after 2023;[16] a more recent study put this date between 2018 and 2020.[17] Attracting foreign investment is thus a critically important issue for the Algerian government, which depended on the hydrocarbon sector accounting for 36 percent of the gross domestic product (GDP), 60 percent of budget revenues, and over 97 percent export earnings in 2010.[18]

Nevertheless, other observers are more optimistic, noting new production projects. Sonatrach made 27 oil and gas discoveries in 2010, while Gazprom and Germany's Ruhrgas made one discovery each. Thirteen discoveries were in the Berkine Basin in eastern Algeria, 10 in the Illizi Basin, 4 in the central region of the Amguid Massaoud Basin, and 2 in the Oued Mya Basin.[19] At the end of 2012, Repsol announced a new gas find at the Tihalatine South well in the Illizi Basin, a new find at the Sud-Est Illizi block in April 2013,[20] and plans four new wells.[21] Repsol also has a major presence in Tinfouye, Tabenkort, and Tifernine. The Menzel Ledjmet East project led by Italy's ENI started production in February 2013.[22] The UK's BP and Norway's Statoil have a joint venture in two fields, including Salah Gas, which is expected to start production later in 2013.[23] The U.S. firm Hess (49 percent partner with Sonatrach) is investing in the Gassi El Agreb

Redevelopment Project to boost production in three long-standing oil fields.[24] Sonatrach is also banking on production of 100,000 bpd at the El Merk oil hub (with partner Anadarko and where production started in May 2013[25]); on Bir Seba (PetroVietnam,[26] 36,000 bpd); on Mezel Ledjmet (formerly ConocoPhillips, 300,000 bpd); and on the Illizi Basin in the southeast (Russia's Rosneft, 38,000 bpd).[27]

China's CNOOC, along with partner Sonatrach and Thailand's PTTEP, successfully discovered crude oil in September 2012 in the fourth and fifth exploration wells in the Hassi Bir Rekaiz project.[28] China's CNPC has acquired the exploration licenses for Blocks 102a/112 in the Chelif Basin in northern Algeria (75 percent operator; 25 percent Sonatrach), Block 350 of the northwestern Oued Mya Basin in the north of the Sahara Desert (75 percent operator; 25 percent Sonatrach), and Block 438b, also in the Oued Mya Basin (100 percent).[29]

The Algerian government which has a 5-year, 2012-16 plan to invest $80 billion in its energy sector,[30] has been constructing new liquefied natural gas (LNG) facilities and intends to increase its exports of natural gas to 100 bcm by 2015, up 60 percent from 2012 levels.[31] Algeria is in the process of developing its Southwest Gas Project, which includes the 102 billion cubic feet per year (bcf/y) Reggane Nord fields led by Spain's Repsol (expected production of 8 million cubic meters per day (m3/d) by mid-2016), the 56 bcf/y Timimoun project led by France's Total (with production starting in 2016),[32] and the 159 bcf/y Touat project led by France's GDF Suez.[33]

A Russian joint venture of Rosneft and Stroytransfaz (jointly holding 60 percent, with Sonatrach holding 40 percent) operates Block 245-S in southern Algeria.[34] The partners are delineating the reservoir at three previously made discoveries in east and west Takazet and north Tisselit.[35]

Angola.

Although Angola became an OPEC member in 1997, it has not always followed the organization's production demands. In 2009, for example, Angola's OPEC production target was 1.51 million bpd, but the country actually produced 1.8 million bpd. Technical and maintenance problems have plagued some of Angola's deepwater fields for years and will continue to limit Angola's crude oil production in the short term. Nonetheless, the International Energy Agency (IEA) still anticipates Angolan crude oil output to increase gradually over the next 2 years as new deepwater production more than offsets chronic maintenance-related declines.[36] The government, which depended on oil exports for 98 percent of government revenues in 2011, plans to boost oil and natural gas production over the coming decade.[37] Crude oil accounted for 46 percent of Angola's GDP and 94 percent of exports in 2011. The United States and China are Angola's main markets for crude oil.[38]

Crude Oil.

Angola had proved reserves of 9.5 billion barrels of crude oil, the second-largest in sub-Saharan Africa after Nigeria and 18th in the world. Recent exploration suggests that Angola's reserves may be larger than originally estimated, and foreign investors are also beginning to consider that some onshore opportunities may be economically viable. BP put Angola's recoverable reserves at 13.5 billion barrels at the end of 2011.[39]

111

LNG.

Angola had proved reserves of natural gas of 10.95 trillion cubic feet (tcf), the fifth-largest in Africa. While 91 percent of Angolan natural gas was re-injected into the country's oil fields in 2011 to aid recovery—or simply flared off—Angola is making efforts to produce and market its natural gas, starting with the Chevron-led Soyo plant, which has a capacity of 5.2 million tons per year of LNG,[40] with start-up planned for June 2013.[41] While gas production had originally been scheduled for shipment to Mississippi, the shale gas boom in the United States has made this unfavorable, and Angola is now targeting consumers in Europe and Asia.[42]

Angola has three main sedimentary basins: the Congo to the north, also straddling offshore and onshore the coasts of the Democratic Republic of Congo (DRC) and Congo (Brazzaville); Kwanza, straddling the central coastal region; and Namibe, straddling the coast in the south and extending into Namibia. All three basins have the pre-requisites for hydrocarbons, but Namibe has not yet yielded them in commercial quantities.[43] In 1955, the first onshore commercial discovery was made at the Benfica Field in the Kwanza Basin. In 1966, the first offshore commercial discovery was made at the Limba Field in the Lower Congo Basin. During much of the 27-year civil war starting at independence from Portugal in 1975 and ending in 2002, exploration activity onshore was restricted. However, deepwater exploration began in 1993, leading in 1996 to the first deepwater commercial discovery, the huge Girassol Field in the Lower Congo Basin.[44]

Successful exploration in Angola's pre-salt formations continues to drive optimistic oil production forecasts for the country. Recently, there have been new discoveries made on several oil fields in Angola, including at offshore fields Greater Plutonia, PSVM, Girassol, Pazflor, CLOV, Chissonga, and onshore in Cabinda South.[45] Oil exploration and production in Angola mainly occur offshore, divided into three bands: Band A, with shallow water Blocks 0-13; Band B with deepwater Blocks 14 to 30; and Band C with ultra-deepwater Blocks 31-40.[46] For example, Italy's ENI made its ninth discovery in deepwater Block 15/06 in March 2013.[47] Oil majors spent $3.55 billion acquiring acreage in offshore Kwanza Basin in 2011. Spain's Repsol has 15 licenses, three of which were awarded in December 2011 for Kwanza.[48] This basin is said to hold the most potential because, 122 million years ago, it adjoined the Campos Basin, which is Brazil's largest petroleum province.[49]

Angola will remain the top destination in Africa and the Middle East for offshore drilling over the next few years. One consultancy estimated that expenditures on offshore drilling would rise to $6.67 billion in 2016, with Nigeria and Egypt a distant second and third at $2.26 billion and $1.52 billion respectively.[50] Although crude oil production in Angola slipped to 1.79 million bpd in 2011, the government is targeting 2 million bpd production levels by 2014.

The large majority of current production is made offshore by three majors: France's Total, with 648,200 operated bpd, U.S. firm Chevron with 434,000 operated bpd, and ExxonMobil with 394,300 operated bpd.[51] Total's production mainly comes from Blocks 0 (offshore Cabinda) and Block 17 (which has three producing zones, as well as Girassol, Dali, and Pazflo. Another field, CLOV, is expected to start in 2014[52] and reach production of 160,000 bpd.[53] For Block 17, Total is operator with a 40 percent share along with Norway's Statoil (23.33 percent), [54] ExxonMobil (20 percent),[55] and BP (16.67 percent).[56]

Chevron operates Block 0 in the Congo Basin in Cabinda, where the mighty Congo River deposited huge amounts of sediment along with vegetative matter that eventually turned into oil over geological time.[57] Although some offshore fields in Cabinda enclave are aging, others are coming online. In February 2013, Chevron approved the $5.6 billion Mafumeira Sul project in Block 0 (Cabinda), including 50 wells and 75 miles of subsea pipelines. The project, for which Chevron as operator will have a 39.2 percent share and partner with Sonagol, Total, and ENI, is expected to see first oil in 2015 and build toward peak total daily production of 110,000 bpd and 10,000 barrels of liquid petroleum gas (LPG) commercialized through the Angola LNG plant in Soyo. Mafumeira Norte, the initial, northern development of this field, achieved first oil in 2009 and currently produces more than 40,000 bpd.[58] Other than at Block 0, Chevron has three other concessions in Angola: Block 2 offshore northwest Angola, Block 14 in deepwater, and onshore Fina Sonagol Texaco.[59]

As noted earlier, China's Sinopec and CNPC hold non-operator shares in several offshore blocks in Angola, the most important of which may be joint venture China Sonangol's 27.5 percent share in Block 17/06 and 20 percent share in Block 32.[60] China Sonagol also has non-operator interests in six other oil blocks in Angola (3/05, 3/05A, 15/06, 18/06, 18, and 31).[61] U.S. firm Marathon owns a 10 percent interest in Blocks 31 and 32.[62] U.S. firm ConocoPhillips acquired, in January 2012, 30 percent operator interests in two deep-water offshore licenses for Blocks 36 (with partners Sonagol, 50 percent, and China Sonagol, 20 percent) and 37 (partners Sonangol 50 percent and Spain's Repsol 20 percent). ConocoPhillips expects first drilling in these blocks in 2013-14.[63]

A number of other smaller IOCs are also active in Angola. Denmark's Maersk Oil made a discovery in offshore Congo Basin Block 16 in September 2012.[64] U.S. firm Cobalt International has sought pre-salt properties in West Africa, and has acquired operator rights to Angola's offshore blocks nearshore 9, and deepwater 20 and 21.[65] Cobalt has a four-well exploratory well program starting in the Lontra Prospect in Block 20,[66] with the first well to spud in 2013 2Q. Cobalt's two exploratory wells in Block 21 confirmed the presence of a mound reservoir, but failed to flow oil.[67]

Onshore.

The government plans to open another licensing round in 2013 for the country's onshore blocks in the pre-salt Kwanza Basin[68] and the Lower Congo Basin.[69] Fifteen blocks will reportedly be offered, all located within close distance of the Atlantic shore rather than far into the interior. Most of Angola's earliest oil discoveries were made onshore in the 1950s, but remained lightly explored due to security concerns that decades later are finally receding. At present, Italy's ENI plans to drill two exploration wells in Cabinda's onshore North block starting in mid-2013.[70]

Shared Development Areas: Progress with Congo,
Agreement with DRC Soon?

In March 2012, Angola and Congo (Brazzaville) signed an agreement to jointly exploit the Lianzi cross-border oil field on their mutual border and share profits equally.[71] Following this, Chevron announced in July 2012 that it would proceed with five partners, including Angola and Congo National Oil Companies (NOCs), in the development of the Lianzi Field, via a tieback to its existing BBLT platform in Angola's Block 14.[72] The boundary dispute with the DRC is more intractable, but observers hope the Lianzi deal will inspire an agreement on shared development. In September 2012, the DRC Minister of Hydrocarbons predicted that the two countries would come to an agreement over the so-called Common Interest Zone in 6 months, but that shared production of any resources was still some time off.[73]

Benin.

Benin's estimated oil and gas reserves are a modest 4-8 million barrels, and current interest by Western majors remains limited.[74] Sillenger Exploration, which has a contract with Benin's government to attract foreign investment in natural resources, estimated in September 2011 that Benin had five billion barrels of oil — i.e., roughly 1,000 times greater than proven reserves.[75] Benin's offshore sedimentary basin, also known as the Dahomey Basin, is part of the larger Transform Margin of the Gulf of Guinea extending from the Keta Basin in eastern offshore Ghana eastward through Togo and Benin to the Mont Okitipupa Basin in Nigeria.[76] The Dahomey Basin in Benin also includes the Hihon-1 and Fifa-1 discoveries (drilled by Kerr-McGee, which was later acquired by Anadarko).

Benin's government has designated 12 offshore blocks, but the only producing field, Seme, in Block 1, produced a maximum of 8,000 bpd in 1982 tapered down to 1,900 bpd,[77] and was closed and restarted by multiple operators.[78] The latest of these operators is Nigeria's South Atlantic Petroleum (SAPETRO), which plans to drill three production wells in 2013 4Q to produce a peak of 6,000 bpd and drain the field.[79] Addax Petroleum, a Canadian subsidiary of China's Sinopec, has a 50 percent interest in Block 1.[80] The UK's Century International is attempting to farm out its interest in Block 2, citing estimates of four billion bbls OOIP in the Gbenonko East and West prospects.[81] In 2005, the Korea National Oil Corporation had interests in two offshore blocks and drilled in Block 2. South Africa's Signet acquired a 90 percent interest in Block 3 in 2011.[82] Brazil's Petrobras acquired in 2011 a 50 percent share of offshore block 4; the remainder is owned by Brazil's Lusitania Petroleum. If 3D seismic confirms potential, this Brazilian joint venture plans three exploratory wells.[83] After commissioning 3D seismic data in 2011, Nigeria's Oranto is seeking an experienced deepwater operator to farm-in part of its 100 percent interest in deepwater Blocks 5 and 6. Oranto also has a 100 percent interest in Block 8 which, like Blocks 5 and 6, it acquired after direct negotiations with the Benin government. Onshore, a tar belt in Benin's coastal region has been sourced to an offshore source of hydrocarbons, but is apparently not being actively explored.[84]

Botswana.

Botswana has an unknown quantity of natural gas deposits in the Kalahari Karoo Basin and the Kgalagadi (Kalahari) Desert. Current exploration efforts largely focus on coal-bed methane.[85] A joint venture between Australia's Origin Energy and South Africa's Sasol began in June 2012 the first part of a three-year program to explore for coal-bed methane in three licensed areas.[86] The program also included an extensive drilling program. Origin's CEO Grant King said that "The delineation of a large [coal-bed methane] resource in Botswana could be used for power generation and liquid fuels production in southern Africa."[87] A second firm, Australia's Sunbird Energy, has eight permits for exploration in the Kalahari Karoo Basin in northern Botswana, approximately 100 kilometers (km) south of the Wankie coalfield in Zimbabwe. The firm planned an aeromagnetic survey in 2012 followed by a two core-hole drilling program.[88] Australian junior Tlou Energy set a stock listing in February 2013 to raise funds to a coal-bed methane appraisal program in Botswana.[89] Australia's Magnum Gas & Power has prospecting licenses in the Nata region over the Ngwasha Basin and the Central Project over the Mmashoro Basin.[90] (See also sections on Mozambique, Namibia, South Africa, and Zimbabwe for more on coal-bed methane in southern Africa.)

Some oil exploration companies have started to make bets that the East Africa Rift System has a branch that travels through Zambia, Botswana, and northeastern Namibia.[91] South Africa's SacOil obtained three petroleum prospecting permits in Botswana, and Magnum Gas & Power has a petroleum and natural gas exploration license in the Ngamiland and Central districts.[92]

Burkina Faso.

There are no known hydrocarbons in Burkina Faso.

Burundi.

Burundi's Minister of Energy and Mines, Come Manirakiza, announced in September 2012 that hydrocarbons had been found in his nation based on "geochemical and petrochemical analyses of surface samples taken from Lake Tanganyika and the Rusizi plain."[93] Burundi has designated four blocks for oil exploration: onshore Block A in the Rusizu Basin, attributed to Nigeria's A-Z Petroleum, and Blocks B-D running north-to-south in Burundian territorial waters on Lake Tanganyika. Blocks B and D are held by the UK's Surestream Petroleum, while Block C is held by Nigeria's Minergy, which in turn is 87.5 percent owned by South Africa's Signet Petroleum.[94] Burundi's blocks are located in the western branch of the Tanganyika Graben that, in turn, is part of the larger East African Rift System.

Aeromagnetic studies in 1982 suggested that Burundi may have hydrocarbon resources.[95] Some experts view Lake Tanganyika's geology as analogous to the tertiary rift system at Lake Albert to the north between DRC and Uganda, where large oil discoveries have been made.[96] Surestream plans to complete seismic studies in 2014-15, with drilling carried out afterwards.[97]

Cameroon.

Cameroon has two main basins with proven hydrocarbons:

- offshore Rio de Rey, between the northwest edge of Cameroon and Equatorial Guinea's Bioko Island to the south, where 89 percent of commercial oil deposits are located; and
- offshore and onshore in Douala/Kribi-Campo in the southwestern coast, where 11 percent of oil deposits are located.[98]

Rio del Rey is part of the southern margin of the larger Niger Delta Basin.[99] Two other basins with possible hydrocarbons are onshore: the Mamfe Basin in the southwest,[100] and the Logone-Birni Basin in northeast.[101] There has been significant, recent exploration in both the Rio del Rey and the Douala/Kribi-Campo Basin, with discoveries made of oil and/or gas/condensate.[102] In 2010, Cameroon's proven oil reserves were a modest 200 million barrels; its proven natural gas reserves were 4.77 tcf.[103]

While Cameroon's oil production has been declining since 1985, when production peaked at 181,000 bpd,[104] the government announced in January 2012 that it expected production would rise about 59 percent by the end of 2012 because of the development of new fields. The Cameroon government also launched in January 2013 a new licensing round for five onshore and offshore blocks in three areas: Lungahe, Bomana, and Dissoni in the Rio del Rey Basin, Kombe-Nsepe in the Douala/Kribi-Campo Basin, and Manyu in the Mamfe Basin.[105]

A major player in Cameroon is Anglo-French Perenco, which purchased the local upstream assets of France's Total in 2011. At offshore Ebome Marine, Perenco has offset the natural decline of these fields by an aggressive program of satellite developments. Similarly, Perenco is optimizing production at its Moudi field with an active work-over program with ESP, gas lift, and sand control. Perenco is moving forward with its plans to develop the Dissoni Nord discovery made in 2005 and expects to stabilize operated production at 40,000 bpd[106] — a large fraction of Cameroon's average production of 63,000 bpd in 2012.[107] The firm is also pursuing an ambitious exploration program in Kombe-Nsepe.[108]

China's Sinopec, through its Addax subsidiary, has a 37.5 percent share in Dissoni. Addax is also the 100 percent operator of the Iroko license area, located approximately 30 km offshore, the 60 percent operator of the Ngosso license area in Rio del Rey Basin, and the 40 percent operator in Lokele's Mokoko Abana Field. The firm also has 24.5 to 32.25 percent shares in eight fields in Lokele (2) and Rio del Ray (6) operated by Perenco.[109] China's Sinopec bought in November 2011 an 80 percent stake in Shell Oil's Cameroon subsidiary, with the 20 percent remainder owned by NOC SNH. This gave China stakes in 12 offshore blocks in Cameroon.[110]

The UK's Bowleven, acting as operator with a 75 percent share, and Vitol, a Dutch oil trader with a 25 percent share, have PSCs for three shallow offshore blocks, MLHP 5, 6, and 7, collectively known as the Etinde Permit, running northwest to southeast along the coast between the Rio del Rey and Douala Basins and bordering Equatorial Guinea's Bioko Island. From January 2012[111] to February 2013,[112] Bowleven made a series of commercial discoveries during well drilling in MLPH 5 and 7.[113] Bowleven also has 100 percent shares in OLHP 1 and 2, two onshore blocks near Mt. Cameroon in the Bomono Permit Area, which is an onshore extension of the Douala Basin,[114] and was preparing for a first exploratory well in early-2013.

In March 2012, Bowleven was also engaged in discussions with the Cameroon government about the construction of a fertilizer plant using natural gas.[115] Switzerland's Glencore is the 90 percent operator of the Matanda offshore block and plans two exploratory wells for 2013. (Bermuda-based AFEX owns the remaining 10 percent.)[116] Vietnam NOC PetroVietnam has a license for the Bomana offshore block.[117]

In 2013 2Q, the U.S. firm Kosmos drilled its SIPO-1 exploratory well in the Liwenyi Prospect in its 100 percent operated Ndian River Block, which is part of the Niger River Delta abutting the border with Nigeria to the north and bordering the Rio del Rey Basin to the south. However, the firm found evidence of a working petroleum system, but did not encounter commercial reservoirs.[118] The company's second 100 percent operated onshore block, Fako, borders to the southeast of the Ndian River block and straddles the Rio del Rey and Douala/Kribi-Campo Basins.[119] Kosmos acquired the second block in 2012 and will carry out a six-year program including one exploratory well.[120]

The U.S. firm Murphy Oil has operator interests in two offshore blocks, 100 percent in Mer Profonde Nord and 50 percent in Mer Profonde Sud, the latter with the UK's Sterling Energy holding the remaining 50 percent.[121] Murphy drilled an unsuccessful well in Mer Profonde Nord in November 2012,[122] but plans to drill up to two more wells in 2013. In 2011, Murphy also purchased a 50 percent operator interest from Sterling for the Ntem deepwater concession in the Southern Douala/Rio Muni Basin adjacent to the northern maritime border of the Rio Muni province of Equatorial Guinea. (Sterling retains the remaining 50 percent share.) Operations in the Ntem concession are currently suspended under the force majeure provisions of the license owing to an overlapping maritime border claim between Cameroon and Equatorial Guinea, but both countries are working to resolve their dispute.[123]

The U.S. firm Noble Energy has 50 percent operator interests in the offshore YoYo mining concession and Tilapia exploration block,[124] both in the Douala/Kribi-Campo Basin bordering on the coast to the east and Equatorial Guinea's maritime border to the west. (Malaysia's Petronas holds the remaining 50 percent interest.[125]) Noble said these Douala Basin interests gave it "upside potential;"[126] nevertheless the firm drilled a dry hole in October 2012 in Tilapia.[127]

Exploration in Contested Bakassi.

Dana Petroleum, a subsidiary of the Korean National Oil Company (KNOC), is 55 percent share operator of the Bakassi West Block in the northern Rio Del Rey Basin, partnering with two Canadian firms, Madison Petrogas (35 percent share) and SoftRock Oil and Gas (10 percent). [128] The Bakassi region is considered to be rich in oil and gas, but has had no modern exploration due to a past territorial dispute with Nigeria, which was settled in Cameroon's favor by internationally sponsored treaties in 2002 and 2006.[129] As part of the reconciliation between the two countries, the UN Special Envoy for West Africa, the Algerian Said Djinnit, in 2011 called on both countries to jointly develop Bakassi's oil reserves. A senior Nigerian official called on China's Addax Petroleum to play a leading role in joint development since it already operates in both Nigeria and Cameroon,[130] but it remains to be seen how interested Cameroon is in joint development.

Onshore Northeast Cameroon.

China's Yang Chang Logone Development Holding Co. Ltd. acquired two blocks, Makary near Lake Chad, and Zina, about 20 km north of the Waza Park.[131] In January 2013, this Chinese firm temporarily suspended its exploration operations in the blocks, which flank Chad's southern oil wells in Doba, after devastating floods.[132]

Monetizing Natural Gas.

Gas reserves in Cameroon are substantial, but had not been exploited previously because of a limited market. The Cameroon government, eager to monetize its 3.5 tcf in technical gas reserves, began a feasibility study for an LNG plant in 2009, with Gaz de France Suez,[133] whose President predicted that Cameroon could become an important actor in commercial natural gas by 2018.[134] The plant, which would be located 30 km south of Kribi and connected to the national gas pipeline network, would have an annual capacity of 3.5 million tons for its first train and use natural gas that is currently flared.[135] The Cameroon government also wishes to build a second LNG plant at Douala.[136] Perenco signed Cameroon's first gas sales agreement, in which its offshore Sanaga field will supply gas to an onshore power plant.[137] The United Kingdom's (UK) Victoria Oil and Gas is working to commercialize the natural gas reserves on its Logbaba permit by signing approximately 20 gas sales agreements[138] with industrial customers in Douala, including breweries and food processing plants, and anticipates daily sales volumes reaching 12 million standard cubic feet per day (mmscf/d) by December 2013, with an estimate of total gas demand in excess of 100 mmscf/d within 5 years.[139]

Chad-Cameroon Pipeline.

Chad and Cameroon have been exporting oil from the Doba crude oil fields in southern Chad via a 1,000-km pipeline through neighboring Cameroon to the port of Kribi, on the Gulf of Guinea. The pipeline is one of the largest U.S. investments in sub-Saharan Africa, estimated to cost $4.4 billion when it was constructed in 2000, and the majority interest is held by ExxonMobil and Chevron.[140]

Cape Verde.

In 2006, Angola NOC Sonagol had expressed an interest in oil and gas exploration in Cape Verde's archipelago and suggested that oil and gas reserves discovered in neighboring Mauritania could stretch as far as Cape Verde. Cape Verde's Prime Minister Jose Maria Neves indicated in July 2010 that his country was hopeful that oil and gas might be found in deep-water pre-salt layers in its Exclusive Economic Zone (EEZ).[141] Following his signing of an agreement with visiting Brazilian President Luis Inacio Lula in 2010, Petrobras signed an accord for exploration offshore Cape Verde.[142] Chinese and Senegalese entities also expressed an interest in exploration in Cape Verde.[143]

Central African Republic.

In collaboration with Sudanese partners, China's CNPC restarted oil exploration in April 2011 in five potential fields in Doba Dosea Salamat region in northeastern Central African Republic (CAR) near the border with Chad. CAR reportedly has a proven oil potential of one billion barrels, with some possibility of five times this much.[144] In 1993, Western Geophysical reported that there exists a common Chad-Central African Basin that straddles the frontier between CAR and Chad and is associated with the Central African Rift system. Oil was discovered in Chad in the 1960s, and the development of oil fields near the Chadian border with CAR started in 2000. CNPC identified four promising sub-basins: Bagara (between CAR, Sudan, and South Sudan), Doseo (between CAR and Chad), Salamat (CAR), and Doba/Bangor (Chad).[145]

Despite prospects for oil discovery, petroleum development and production in the CAR will likely be held back by the country's political instability. There have been a series of civilian and military governments since independence in 1960, and the government of the President Francois Bozize, who himself came to power in a military coup in 2003, did not fully control the countryside, particularly in the northern and central parts of the country, where several rebel groups and the Lord's Resistance Army operate.[146] Rebel groups drove Bozize from power on March 23, 2013,[147] and elected self-proclaimed President Michel Djotodia as interim head of state.[148]

Chad.

Today, Chad has proven reserves of 1.5 billion barrels with excellent prospects to increase this figure.[149] Chad shares with its neighbors three sedimentary basins known for oil and gas potential:

1) the Koufra–Erdis Basin in the north shared with Libya;

2) the Chad Basin in the center-west shared with Niger, Northern Nigeria and Cameroon; and,

3) the Sud Basin in the south shared with CAR and South Sudan.

Oil exploration began in Chad in the 1950s and led to discoveries in the 1960s before developments were blocked by political instability. In the 1990s, discussions between the World Bank, the Chad and Cameroon governments, and a private consortium led to a decision to build simultaneously in 2000, the "Esso Chad Doba Exploitation" and the Doba-Kribi pipeline.[150] Eighty-five percent of this pipeline is located in Cameroon.[151] Since 2003, ExxonMobil, Chevron, and Malaysia's Petronas, through two consortia in Chad and Cameroon, have been exporting oil from the Doba crude oil fields in southern Chad via a 1,000-km pipeline to Cameroon's Kribi port, on the Gulf of Guinea.

Chad's major oil fields are the Bolobo, Kome, and Miandou in the Doba Basin, which are estimated to contain reserves of 900 million barrels.[152] ExxonMobil's oil production in Chad peaked at 126,200 bpd in 2010,[153] and averaged 115,000 bpd in 2011. (ExxonMobil's operator share in the seven fields in Doba[154] was 40 percent, with Petronas holding 35 percent and Chevron 25 percent.)[155] The company drilled 69 additional production wells in 2011.[156]

New companies, including China's CNPC and Canada's Griffiths Energy International (which is changing its name to Caracal), have also reached agreement with the Chad and Cameroon entities to use the pipeline to export oil from their fields[157] — making the author wonder if there is not a risk of exceeding the pipeline's current capacity. Griffiths hopes to complete as early as 2013 2Q a 111-km connection pipeline to link its Badila and Mangara fields to the export pipeline.[158] Griffiths successfully drilled two exploratory wells — Badila-1 and -2 in May 2012 and January 2013,[159] and commenced drilling of Mangara-4 in March 2013.[160] Griffiths, which sold a 25 percent interest in the fields to Switzerland's Glencore International in September 2012[161] to finance the development of Badila and Mangara,[162] became embroiled in a bribery scandal in January 2013 when Griffiths pled guilty in Canada to having bribed Chadian officials in order to gain its production sharing contracts (PSCs) in 2011.[163]

The U.S. firm ERHC won 100 percent shares in three blocks in 2011, two of which — BDS 2008 and the northern half of Chari Ouest III — fall on the north flank of the Doba/Doseo Basin in southern Chad, where ExxonMobil and others have discovered 1.29 billion BOE.[164] The third block, known as Manga, is in northwest Chad, adjacent to the Niger border.[165] The Canadian firm Simba Energy won 100 percent shares in three blocks in September 2012, two of which — Chari Sud I and the southern half of Chari Sud II, both bordering with the CAR — are also in the Doba/Doseo Basins south of Mangar and Badila, where their large proven reserves are located.[166] The third block, Erdi III, is located in northeastern Chad near the Libya border.[167] The Doba/Dosea Basins are part of the West and Central African Rift Systems that extend across central Africa from Nigeria to Kenya to Sudan.[168]

Earlier, the Chadian government tried to boost the potential of other fields through diplomatic channels. While Chad was one of the last countries in Africa to keep diplomatic ties with Taiwan authorities (and not the Peoples' Republic of China),[169] Taipei's

NOC, China Petroleum Corporation, actively engaged in oil exploration. Even today, this NOC retained an interest in two blocks, BCO III/BCS 11/BLT I,[170] in Chad where it discovered oil in 2011. In 2003, CNPC bought out the shares of a Swiss company for Block H, in whose Bongor Basin the company subsequently made several commercial discoveries starting in 2006. In 2007, CNPC and the Chadian Ministry of Petroleum signed an agreement for the establishment of a 60-40 joint venture, 20,000-bpd refinery in Ndjamena, which started production in 2011.[171] CNPC completed in 2011 the construction of a 311-km pipeline from its fields to the new refinery.[172] Based significantly on the growing weight of PRC investment in Chad, Ndjamena re-established diplomatic relations with Beijing in October 2009.[173]

Comoros.

The Comoros government awarded its first oil exploration license in March 2012 to the privately held Kenyan company Bahari Resources for an offshore region on the western boundary of the Comoros. The acreage is adjacent to offshore Area 1 and Area 4 of Mozambique's Rovuma Delta, where the recent giant hydrocarbon discoveries were made by the U.S. firm Anadarko and Italian oil giant ENI. Under the agreement, Bahari will undertake a phased seismic and drilling program within the licensed area and will carry out for the government a regional study of the Comoros "entire territory." The report will be used for the demarcation of blocks for a future licensing round.[174] Bahari will also fund a government program to develop its internal expertise in upstream petroleum, including management and legislation. The government hopes to send to Parliament by October 2013 a draft petroleum law that will govern all aspects of petroleum exploration and production.[175]

Unfortunately, the March 2012 license award has become mired in controversy as Luxemburg-based Boulle Mining claims that it had signed a prior, exclusive exploration and production agreement with the Comoros government in November 2011. Boulle's subsidiary, Mozambique Channel Discovery, filed for arbitration at the International Chamber of Commerce in Paris in November 2012, contesting the license issued to Bahari.[176] For its part, the Comoros government communicated to Boulle in May 2012 that its earlier agreement was invalid because it was signed by the Finance Minister in an acting capacity and did not receive approval by the Council of Ministers. Instead, the government asserts, only the Vice President and the Ministry for Energy, the Environment, and Industry are "the only mandated authorities in this matter."[177]

As a sign of how feverish interest in Comoros territorial waters has become, former Directors of Cove Energy, which sold out its 10 percent stake in Rovuma Area 1 in Mozambique for $1.9 billion to Thailand's PTTEP, formed a new company, Discover Exploration, specifically to raise capital and attempt to win an exploration license from the Comoros in March 2013.[178] That same month, Discover announced that the Comoros had signed a PSC with Discover and Bahari for a block offshore Comoros adjacent to the block Cove Energy sold in Mozambique.[179]

The U.S. firm GX Technologies completed in May 2011 seismic studies of part of Comoros' territorial waters, from the Tanzanian maritime border at Mtwara, southward to Tanzania's Pemba, then to the island of Mayotte, which is part of France but claimed by the Comoros.[180] In July 2012, the Comoros government signed an agreement with the Norwegian firm TGS Nopec for (seismic) exploration of other zones in the country's territorial waters.[181] The UK's Avana Petroleum has carried out "an initial analysis of the hydrocarbons prospectively of the waters around the Comoros Islands,"[182] but this may mean that it purchased seismic data from GT Exploration or another contractor.

Congo (Brazzaville).

Congo is the fourth-largest oil producer in Africa that is not part of OPEC, after Egypt, the Sudans (counted jointly), and Equatorial Guinea. Congo had proven oil reserves of 1.94 billion barrels at the end of 2011.[183] Congo produced 302,000 bpd in 2010, surpassing the previous peak of 292,000 bpd in 2000. In 2010, 49 percent of Congo's oil exports went to the United States, and 31 percent went to China.[184] Congo has produced oil since the 1960s, with production increasing until about 2000, and declining afterwards until 2008, when new offshore fields reversed this trend. Congo's first deepwater field, Moho-Bilondo, has been the most important. This field, operated by France's Total, came online in 2008 and reached production of 90,000 bpd in 2010. Total's share in the field is 53.5 percent, along with partners Chevron with 31.5 percent, and Congo NOC SNPC with 15 percent. Total operates nine of the 22 active fields in Congo and has a non-operator share in two others.[185] Total has decided to move forward with the $10 billion deepwater Mogo-Bilondo "Phase 1 bis" and Moho North joint development project, with production beginning, respectively in 2015 and 2016,[186] and peak production of

140,000 bpd in 2017.[187] Two appraisal wells drilled by Total in 2010 also confirm the growth potential of the southern part of the field.[188] Italy's ENI produced over one-quarter of Congo's crude oil in 2010, with production hitting 98,000 bpd.[189]

The U.S. firm Murphy Oil is the operator of two offshore blocks, Mer Profonde Nord (100 percent share) and Mer Profonde Sud (50 percent share), where it discovered oil in 2005 and started production in 2010.[190] ExxonMobil tried to sell off its 30 percent, non-operator share in the deepwater Mer Tres Profonde Sud Block in 2010 that is operated by Total (40 percent) in partnership with ENI (30 percent).[191] Anglo-French company Perenco operates four offshore fields with relatively small production.[192] France's Maurel et Prom operates and is exploring the Noumbi onshore field, which is adjacent to the producing M'Boundi onshore fields[193] and shares the same sedimentary basin as Gabon.[194] The UK's SOCO International operates and is exploring the offshore Marine XI field in the North Congo Basin.[195] (The presence of oil in Congo, as in Angola's Cabinda enclave, is explained geologically by sediments carried by the Congo River.[196]) Foreign NOCs and IOCs with non-operator roles in Congo are China's CNOOC, PetroVietnam,[197] the Netherlands' Vitol, Sweden's PA Resources and Svenska Petroleum Exploration, and the UK's Afran and Anglo-Irish Tullow Oil.[198] CNOOC hired a ship to drill two wells with an option for a third in its offshore block in 2013 1H.[199]

Oil Sands.

In 2008, the Congo government awarded ENI two exploration permits to launch a pilot oil sands project located in Tchiktanga and Tchikatanga-Makola[200] in

southern Congo. Congo has estimated recoverable reserves from the project of between 500 million and 2.5
billion barrels. [201] The exploration phase includes col-
lecting data from sounding holes in order to evaluate
the quality, thickness, and distribution of the tar sand.
The small-scale pilot project that might follow would
"exclude open-cast mining and tailing ponds, which
are considered to have a high-risk impact and both of
which are used in oil sand operations in Canada."[202]

Shared Development Areas.

As noted previously in the section on Angola,
Congo (Brazzaville) and Angola signed an agreement
in March 2012 to jointly exploit the Lianzi cross-border oil field on their mutual border and share profits
equally. Following this, Chevron announced in July
2012 that it would proceed with five partners, including Angola and Congo NOCs, in the development of
the Lianzi Field, via a tieback to its existing BBLT platform in Angola's Block 14.

Natural Gas.

Congo also has large reserves of associated gas—
the fifth largest in sub-Saharan Africa at 3.2 tcf. The
country produced 298 bcf of natural gas in 2010, although only 14 percent (43 bcf) was marketed. Sixty-
five percent was re-injected, and the remaining 21
percent was flared or vented.[203] ENI signed an agreement in 2008 to eliminate all gas flaring at its Congo
operations by 2012 by constructing two electric power
stations fueled by associated natural gas from the
M'boundi Field and by re-injecting the remaining gas
into the deposit.[204]

Cote d'Ivoire.

In 2010, crude oil production decreased by an estimated 24 percent to 16.4 million barrels.[205] Cote d'Ivoire has proven reserves of 100 million barrels of oil and 1.1 tcf of natural gas, but a dramatic increase in exploration and drilling after oil and gas finds in neighboring Ghana are likely to significantly boost these figures in coming years. The producing Baobab Field, which came on stream in 2005, has an estimated 200 million barrels of oil and the producing Espoir Field, which came on stream in 2002, is estimated to contain a further 120 million barrels of oil.[206] Baobab has been operated by Canadian Natural Resources (57.6 percent), along with Sweden's Svenska Petroleum Exploration (27.4 percent) and Petroci (15 percent).[207] The Foxtrot offshore gas field was discovered along with the Panthere gas and condensate field in 1993 and the Lion oilfield in 1994, both of which came on stream in 1995. UK firm Afren (47.96 percent share) operates the Lion and Panthere Fields, whose production averaged almost 5,000 bpd of oil equivalents, with SK Energy (12.96 percent share) and two Ivorian partners. In its 65 percent share of Block CI-01, which is on the maritime border with Ghana and west of the Jubilee Field, Afren is carrying out 3D seismic work. Block CI-01 has been the sight of 10 successful discoveries of oil and gas in the Ibex and Kudu Fields, while only gas was discovered in the Eland Field.[208]

Italy's Edison International made a discovery in Block CI-24 in September 2010 in the Baobab Field, which it shares with Sweden's Svenska Petroleum Exploration (20 percent) and Petroci.[209] Russia's Lukoil made an oil discovery in offshore Block CI-401 in December 2011. It operates the block with a 56.66 percent interest, along with U.S. firm PanAtlantic (formerly Vanco) with a 28.34 percent share, and Petroci with 15 percent.[210] Lukoil, which also has a share in Block CI-101,[211] signed a PSA with the Ivoirian government in October 2012 for a 60 percent share of offshore Block CI-524, along with PanAtlantic (30 percent) and Petroci (10 percent).[212] In February 2013, Lukoil announced that it planned to spend $400 million in 2013 to drill more wells near the site of its discovery in Block CI-401, and carry out further exploration on 3-4 offshore blocks.[213] Operator Anglo-Irish Tullow Oil discovered oil in exploration Block CI-103 (share 45 percent), which is east of the Espoir Field, along with partners Anadarko (40 percent) and Cote d'Ivoire NOC Petroci (15 percent share). In June 2012, Tullow's Exploration Director Angus McCoss said: "The discovery of light oil in our first well in CI-103 extends the proven play for oil westwards from our successes in Ghana. . . . We look forward to further drilling in CI-103 during 2013."[214] In September 2012, Australia's Rialto Energy estimated that the initial gas production from its Block CI-202 Gazelle Field 80 km west of Ghana's huge Jubilee Field would produce 40 million cf of gas daily. Operator Rialto has an 85 percent interest in the block, shared with Petroci (15 percent).[215]

France's major, Total, discovered oil at its Ivoire 1X deepwater exploration well in May 2013. The well "confirms the extension into Block CI-100 of the already proved active petroleum system in the prolific Tano Basin, home to several fields, including Jubilee in Ghana."[216] In 2010, Total acquired a 60 percent interest in Block CI-100, with Yam's Petroleum[217] owning 25 percent and Petroci 15 percent.[218] One observer believes that Cote d'Ivoire may be more willing to settle a maritime boundary dispute with Ghana now that there is a find that is definitively on its territory and pressure from major oil companies.[219] In February 2013, Total secured three new "ultra-deep" offshore licenses, as operator (54 percent share) of CI-514 with Canadian Natural Resources (36 percent) and Petroci (10 percent), and as operator (45 percent share) of CI-515 and CI-516 with Anadarko (45 percent) and Petroci (10 percent). Total's Senior Vice President for Exploration, Marc Blaizot explained the acquisitions as follows:

> The so-called Abrupt Margin theme that will be exploring in this acreage is . . . the same theme in exploration licenses in French Guyana, where a promising discovery has already been made, and in Mauritania.[220]

Australia's African Petroleum operates 90 percent shares of Blocks CI-509 and CI-513 with Petroci (10 percent) and hopes to drill one well on each block in late 2013 or early 2014,[221] while the Netherland's Vitol is operator (36 percent share) with Genel Energy (24 percent) and Petroci (40 percent) in offshore Block CI-508, where it plans a first exploratory well in 2014.[222]

The Democratic Republic of the Congo.

Exploration for oil and gas in the DRC from the 1960s to 1983 led to the discovery and development of five oil fields and one still-undeveloped gas field. Of these, the Mbale offshore field contains 48 percent of the Coastal Basin's recoverable reserves.[223] Nevertheless, the future of oil exploration for the DRC lies principally in the east, in the Albertine Graben, where oil has already been discovered in significant quantities across the border in Uganda. The five blocks on the DRC side of Lake Albert are said to hold 800 million barrels of oil.[224]

Unfortunately, controversy has dogged onshore oil Blocks 1 and 2 in eastern DRC. Anglo-Irish Tullow Oil and the UK's Heritage Oil[225] were awarded the blocks in 2006 but the DRC government cancelled the licenses under controversial circumstances, claiming that the contract did not follow international standards. In 2008, South Africa's Divine Inspiration Group signed a competing contract for Block 1. Tullow initially launched legal proceedings, but then abandoned them as unenforceable. In 2010, two other companies from the UK, Caprikat and Foxwhelp, obtained licenses for Blocks 1 and 2 after the government again cancelled the existing licenses.[226] France's Total purchased a now 66.67 percent operator share of Block III in eastern DRC, in the Albertine Graben, with CoHydro owning 15 percent, and South Africa's SacOil 12.47 percent and Divine Inspiration Group 5.86 percent.[227] Sweden's PetroAfrican Resources presented in 2011 an extensive geological report of Block IV in the eastern Congolese Rift Zone, concluding that "the Block is a promising area for hydrocarbons."[228] The UK's SOCO's has encountered different challenges in exploiting its 85 percent operator share of Block V in onshore eastern DRC, adjacent to the border with Uganda, with DRC NOC CoHydro holding 15 percent. The block is 7500 km², encompassing a section of the mountain gorilla-rich Virunga National park, including part of Lake Albert,[229] and its exploration has come under criticism by some international environmental groups, including the World Wide Fund for Nature.[230] The European Commission plans to do an environmental study of the area covered by Blocks III (Total-operated), IV (open), and V (SOCO-operated), including evaluating the political risk of the development of oil resources in an area that has seen numerous armed conflicts since the 1990s.[231] Perhaps in part due to such difficulties, the DRC government announced in February 2013 that it planned to adopt as early as April 2013 a new law regulating its oil and gas sector, which would include a provision requiring all potential investors to go through a tender process.[232]

Further south, Canada's Alberta Oilsands concluded a farm-in deal with operator Pan African Oil to jointly develop the Klembe Block (Block 5) and the Fatuma Block (Block 6) in the Kalemie sub-basin on Lake Tanganyika (DRC side), in the heart of the Western Rift Valley of the East African Rift System. These blocks are adjacent to acreage held by Total (in Tanzania on the eastern side of the lake).[233]

Further to the west, SOCO International undertook 2D seismic work at its 65 percent operated share in the Nganzi onshore block in the western Bas-Congo Region near the coast[234] and was examining plans as of March 2013 to drill two wells in the block.[235] Its partners were DRC NOC CoHydro (15 percent) and Japan's INPEX (20 percent). The U.S. firm EnerGulf is the operator of the Lotshi Block in the onshore salt Coastal Basin, which is contiguous with the highly prospective Cabinda enclave of Angola to the north, and in trend with the M'Boundi field in the Republic of Congo in a similar geological graben. This basin also includes the East-Mibale and other fields operated by the Anglo-French Perenco Oil and contiguous to the east with a bitumen mining license with known surface oil seeps.[236] EnerGulf had planned a three-well drilling program in summer 2011,[237] but this was delayed and its DRC partner, CoHydro, now expects that at least one well will be drilled in 2013.[238] A 2011 assessment by an outside consultant reached an estimate of 313 million barrel mean estimate of gross prospective oil resources in the EnerGulf's seven prospects in the Lotshi Block.[239] The Anglo-French company Perenco, currently the DRC's sole oil producer at 28,000 bpd, operates 12 onshore and offshore fields south of Angola's Cabinda and north of Angola's Soyo, storing crude on board the Kalamu, a floating terminal. It has a 50 percent share, with partners INPEX (Teikoku) (32.28 percent) and Chevron (17.72 percent).[240]

Negotiations with Angola.

In September 2012, the DRC Minister of Hydrocarbons predicted that in 6 months, the two countries would come to an agreement over the so-called Common Interest Zone created in 2004, but that shared production of any resources was still some time off. The DRC could receive royalties from production already underway in the Zone.[241] In 2010, U.S. firm Chevron signed an agreement with the DRC government, allowing it to ship gas from Angola's Cabinda enclave, cutting across the DRC's narrow strip of Atlantic coast to its new Soyo LNG plant. The Soyo facility could also be used to process DRC gas, and Chevron has expressed an interest in looking at potential gas deposits in Congo,[242] and in processing DRC gas that is currently flared.[243]

Djibouti.

In 2011, the UK's Oyster Oil and Gas won a 100 percent operator interest in three onshore blocks in the Guban Basin (Blocks 1 and 4), the Red Sea Basin

(Block 2), and one offshore in the Red Sea (Block 3).[244] The Guban Basin extends from Djibouti into Somalia's northern province of Somaliland where there have been oil seeps and recovery of oil from wells drilled in previous decades. The Red Sea Basin has had several discoveries in its northern area in Sudan, Egypt and Saudi Arabia, and there are also analogues with the large Jurassic oil fields located nearby in Yemen.[245] Oyster is confident that oil-bearing basins were formed with the break-up of Gondwanaland in Guban Basin and the Red Sea. Block 1 is mainly onshore, but also extends into the Gulf of Aden Margin.[246] After a July 2012 with Djibouti's Energy Minister, Oyster Chief Executive Officer Ravindra Shaji said that:

> Four months after the commencement of work on different sites in the north and south of Djibouti territory, we have come to report . . . the preliminary finding of this first phase of oil and gas exploration, which we can say are rather encouraging at several sites.[247]

Egypt.

Egypt is the largest oil producer in Africa that is not an OPEC member, and the second largest natural gas producer on the continent after Algeria. Egypt is also the only African oil and gas producer besides Tunisia that consumes domestically all or a large fraction of its production. As of January 2012, Egypt's proven oil reserves were 4.4 billion barrels. In 2011, Egypt's total oil production averaged 710,000,[248] of which approximately 560,000 was crude oil including lease condensates and the remainder natural gas liquids. Egypt's production peaked at over 900,000 bpd in the 1990s. However ongoing successful exploration and enhanced oil recovery techniques in existing fields have eased the decline at aging fields, and boosted proven oil reserves.[249] In the second half of 2012, Egypt made 29 oil and natural gas discoveries; in the first quarter of 2013, there were 13 oil discoveries and four gas discoveries in the Western Desert, the Gulf of Suez, the Mediterranean, and Upper Egypt that added 12 million barrels of crude oil and condensates and 125 bcf of gas to Egypt's reserves.[250]

One of Egypt's challenges has been to satisfy increased domestic demand despite falling domestic production. Domestic oil consumption grew over 30 percent over the last decade, from 550,000 bpd in 2000 to 815,000 bpd in 2011, outpacing production as of 2008. Political unrest from the "Arab spring" and its aftermath did not directly affect foreign oil investors in Egypt as the top producing regions — the offshore Nile Delta and the Western Desert — are far from population centers. Since 2000, oil output in the Western Desert has doubled, and it now accounts for around 28-30 percent of total oil production.[251] Four other major producing areas in Egypt are the Eastern Desert, Mediterranean Sea (onshore and offshore), the Gulf of Suez, and the Sinai Peninsula.[252]

Egypt's natural gas sector has been expanding rapidly, with production more than tripling from 646 bcf in 2000 to 2.2 tcf in 2010. Egypt's proven gas reserves registered at 77 tcf, a sharp increase from 2010 estimates of 58.5 tcf, and the third highest in Africa after Algeria and Nigeria.[253] Three-fourths of Egypt's natural gas production came from Mediterranean offshore blocks, where 78 percent of the country's gas reserves are located.[254] Dry natural gas exports began in 2003 and had been rising rapidly with the completion of the Arab Gas Pipeline in 2004 and the start-up of the first three LNG trains at Damietta in 2005. ENI operates

141

Damietta, and Spain's Repsol is a partner.[255] However, gas exports leveled off in 2006 and fell in 2010, in part due to low international prices and growing domestic demand, which led the government to enact a two-year moratorium on new gas export deals.[256] Electricity shortages during summer 2010 also led to domestic pressure to reduce natural gas exports, which fell to 30 percent of contracted quantities in 2011, in part also due to a series of attacks on a pipeline supplying Israel and Jordan.[257]

The Egyptian government encourages IOCs to participate in the oil and gas sector, and currently more than 50 are operating in Egypt.[258] In April 2013, Egypt announced the award of eight new oil and gas blocks to six companies (out of 15 blocks offered): BP, Dana Gas, the UK-based subsidiary of the Korea National Oil Company, Italy's Edison, Ireland's Petroceltic International, Australia's Pura Vida, and Canada's Sea Dragon Energy.[259] One possible policy change apparently being floated as a trial balloon by the Egyptian government is to change its licensing policy so that future oil and gas contracts have terms more similar to those offered by Algeria and Libya, with a higher take of production for the government—a strategy that risks backfiring as foreign investors are already alarmed about security in North Africa and a rising cost of doing business.[260]

The U.S. firm Apache is the largest producer of liquid hydrocarbons and natural gas in the Western Desert and the third largest in Egypt. In 2011, the firm drilled 221 wells, and produced daily 217,000 bpd of crude oil and 865 mcf of natural gas (including joint venture partner-shares). Apache planned to drill 270 wells in 2013, of which 60 would be exploration wells.[261] Eighty-two percent of its acreage is undeveloped, which includes seven new development leases in the Faghur Basin,[262] providing significant future E&P opportunities.[263] In May 2013, Apache announced finds in new discoveries in three concessions, each in different basins: North Ras Qattara, Siwa, and North Tarek.[264] NOC Oil India has a 25 percent non-operator interest in offshore Blocks 3 (South Quseir in the Red Sea) and 4 (South Sinai, at junction Red Sea/Gulf of Suez).[265]

The UK's BP and its partners produce more than 35 percent of Egypt's natural gas and almost 40 percent of its oil production.[266] BP and Germany's RWE announced a joint investment of $9 billion for the development of the North Alexandria and West Mediterranean deepwater gas fields, owned 60 percent by BP (operator) and 40 percent by RWE.[267] Production at the project was to have started in 2014, but has been delayed by public protests. The project has the potential to unlock around 50 tcf of gas resources in the Nile Delta Basin, which would double Egypt's current gas reserves.[268] Another BP partner, Italy's ENI, claims to be the "largest foreign energy player in the country," and had oil and natural gas equity production in 2011 of 240,000 BOE/D. In February 2013, ENI made a new oil discovery in the Meleiha Concession in the Western Desert.[269] Other major foreign partners of BP in Egypt are Malaysia's Petronas, which shares equal 38 percent and 35.5 percent shares in two 3.6 million tons per annum (mpta) LNG trains and 30 and 50 percent shares in Nile Delta offshore fields; Italy's Edison, which has a 20 percent share in a Nile Delta gas field; and Dana Petroleum, which has a 50 percent share in a Nile Delta offshore field.[270] In 2012, Shell farmed in as operator of the Alam El Shawish West Gas Development Project (40 percent) in the Western Desert with partners Vegas (35 percent) and GSF Suez (25 percent), where 200 mcf/d in gas production is anticipated.[271] The United Arab Emirates' Dana Gas started gas production at two new wells in the Nile Delta in March 2013 and plans to start production at a third well there—all 2012 discoveries—in 2013 2Q, adding 20 million cubic feet daily.[272]

Shale Oil.

Oil shale resources were discovered in the Safaga-Quseir area of the Eastern Desert in the 1940s, where estimated reserve equivalents of in-place shale oil are 4.5 million barrels. In the Abu Tartour area of the Western Desert are estimated to have about 1.2 million barrels of in-place shale oil, while oil shale in the Red Sea area would be extracted by underground mining.[273]

Equatorial Guinea.

Equatorial Guinea produced an average of 252-282,000 barrels of crude per day in 2011 and 243 bcf of natural gas.[274] The country has three major producing fields: the Zafiro oil field, which is the largest in the country and northwest of the island of Bioko; the Alba gas field, to the east of Bioko; and the Ceiba oil field in offshore Block G of the Rio Mini Basin.[275] ExxonMobil has a 71.25 percent interest in Block B,[276] whose Zafiro Field produced 15 percent less in 2011 than the prior year.[277] The U.S. firm Hess has an 80.75 percent operator interest in the Ceiba Field with partners Anglo-Irish Tullow Oil[278] and NOC GEPetrol,[279] and reported a 21 percent decrease in production in 2011.[280] Crude oil production has been declining as a result of maturing oil fields, but the recent start-up of the Aseng oil and gas-condensate field and the anticipated start-up of the Alen gas-condensate field in late 2013 are projected to revive liquids production in the near term.[281] At the end of 2011, Equatorial Guinea had proven reserves of 1.1-1.7 billion barrels,[282] and 1.3 tcf of proven natural gas reserves — the majority of which are found offshore Bioko Island.[283]

Offshore.

The offshore areas of petroleum exploration and development are in the shelf around Bioko Island, which is part of the Niger Basin surrounded by Nigeria, Cameroon, Sao Tome and Principe, and the shelf off Rio Muni, an enclave between Cameroon and Gabon.[284]

Bioko. In September 2012, the UK's Ophir announced that its sixth discovery alone in offshore Block R in the southeastern Niger Delta, southwest of Bioko Island, had added 1 tcf of recoverable gas (80 percent share, remainder with GE Petrol).[285] The firm's multi-well drilling campaign targeted sufficient resources to meet a two train LNG threshold of approximately 2.5 tcf.[286] Southeast of Bioko Island, U.S. firm Noble Energy made a series of oil and gas discoveries in its Block I (38 percent operator),[287] including the 120,000 bpd Aseng oil field,[288] and Block O (45 percent operator), including the Alen gas-condensate field which included the firm's adjacent Yoyo license in Cameroon, total 127 mmbl liquids and 1.25 tcf in natural gas. The firm is considering options for monetizing these finds (including possibly another LNG train).[289] Noble's primary foreign partner in Block I and O are Switzerland's Glencore, with respective shares of 23.8 percent and 25 percent.[290] Glencore also has a 50 percent working share in Block V, southwest of Bioko Island on the maritime frontier with Sao Tome and Principe, partnering with Bermuda-based AFEX Global, which believes the block contains a south-westerly extension of the Cameroon Transpressional Zone that has produced oil and gas discoveries to the northeast.[291] In January 2012, the government issued a press release in which it expressed the desire to see partners in Blocks I, O, and R work together to plan for LNG Train 2,[292] doubling the 3.7 million metric tons per year (mmt/y) Punta Europa LNG Train 1, which started operations on Bioko Island in 2007.[293] In June 2012, the government announced the signature of a production sharing contract for Block A-12 offshore Bioko with the U.S. firm Marathon.[294] Marathon has a 63 percent share (65 percent with government carry interest in the Alba Field, a 52 percent interest in the Alba liquid petroleum gas (LPG) plant, a 45 percent interest in the Atlantic Methanol Production Company (AMPCO), and a 60 percent interest in an LNG facility on Bioko Island.[295] Spain's Repsol is the operator in exploratory Block C-1 next to Bioko Island.[296]

Rio Muni. In the Rio Muni Basin, the UK's White Rose Energy (46.31 percent operator) had planned to start a drilling campaign in 2012 4Q in Block H;[297] U.S. firm PanAtlantic (80 percent operator) has seismic data and is preparing for drilling in Block K—Corisco Deep;[298] China's CNPC (70 percent operator) has been exploring in offshore Block M through its Santa Isabel Petroleum subsidiary.[299] China's CNOOC also has rights to Block S.[300] The government also approached China's Sinopec about constructing a 20,000 bpd refinery.[301]) In December 2012, the government granted

Hong Kong-registered Xuan Energy a license as operator of newly created Block Y, offshore Rio Muni, with a 15 percent participating interest for Brenham Oil & Gas, a subsidiary of American International Industries.[302] Of the seven other production-sharing contracts announced at the same time, two others were given to U.S. firms: Block W, offshore Rio Muni, granted to U.S. firms Murphy Oil (operator) and Pan Atlantic Oil and Gas, with NOC GEPetrol as partner; and Block EG-02 (offshore Bioko Island), to Pan Atlantic Oil and Gas, Novamark International, Atlas Petroleum, and GEPetrol.[303] Murphy Oil was also in negotiations with the government about acquiring interests in offshore Blocks J-14, J-15, K-14 and K-15, along with partners Vanco (US), Dana Petroleum (Korea), and NOC GEPetrol.[304]

Onshore.

In November 2011, Vice Minister of Mines, Industry and Energy, Gabriel Lima, indicated that the government:

> was also inviting companies to participate onshore. We have only one company, Total, which has drilled two wells onshore . . . if you look at the basins onshore in Cameroon and Gabon, that is where they have the majority of the oil.[305]

Maritime Boundary Disputes.

Since mid-1999, there have been strained regional relations over maritime boundaries, particularly the Zafiro Field Block B, which Nigeria has claimed is part of the same structure as its Ekanga oil field, just 3.5 km north of Block B. After Equatorial Guinea's

President Obiang accepted an equidistant median line defining territorial boundaries as stipulated under UNCLOS, Cameroon, Sao Tome and Principe, and Nigeria welcomed this decision. Equatorial Guinea and Gabon have disputed the ownership of three islands in the Gulf of Guinea since the 1970s, but agreed in 2004 to allow joint oil exploration in the disputed territories until a final resolution is worked out under UN mediation.[306]

Eritrea.

Eritrea has been underexplored for oil and gas in part due to political instability, including its war of independence from Ethiopia (1961-91), war with Ethiopia (1998-2000), and an autocratic regime that, more recently, was subjected to UN sanctions in December 2011 requiring that foreign mining (but not explicitly energy) companies ensure funds from the sector are not used to destabilize the region.[307] This being said, there are strong signs that Eritrea has great potential over the long term as an oil and gas producer.[308]

The Eritrean government signed oil exploration agreements in 2008 for the two most northerly offshore blocks on the maritime border with Sudan — Bahri and Defnin — with Defba Oil Share Company, which was reportedly a joint venture between a Chinese company and the Energy Alliance Company, which reportedly had Saudi and Bahrain capital.[309] Canada's Centric Gas and Oil Company signed an exploration agreement with Eritrea in 2010 for a mixed onshore-offshore block in the Dahlak Block in the north-central Dismin region in northeastern Eritrea. Centric's CEO Alec Robinson noted that, historically:

Eight . . . offshore wells had good oil or gas shows, and the widespread occurrence of seeps is further evidence of a working petroleum system. . . . In fact, one well, C1, drilled in 1969 by Mobil, suffered a massive gas blowout, and continued flowing for 55 days before stopping naturally. From the limited . . . data available, a number of prospects and leads have been identified in both the pre- and sub-salt formations of the Eritrean Red Sea.[310]

Eritrea has 12 oil blocks.[311] There has been little public information about the Defba or Centric deals since they were announced, although one March 2013 news item claims that Eritrean President Isaias Afwerki told the Middle East newspaper "Asharq Al-Awsat" that Eritrea has "a large [oil] reserve," but we should not talk about this without verifying all the angles."[312]

The southern part of the Red Sea, where Eritrea is located, is a boundary area between the African and Arabia Plates. Uganda's Basin, which is located in the same East African Basin, has had major oil discoveries in a sedimentary basin only one-quarter of the areal distribution of Eritrea's. U.S. firm Anadarko did promising seismic work in the Zula Block in central offshore Eritrea, but swapped the property with Italy's ENI. Just beyond Eritrea's maritime border with Sudan, IPC's Suakin-2, drilled in the mid-1990s, found gas. Anadarko also drilled three unsuccessful wells in the Edd Block, and withdrew with U.S. partner Burlington Resources from the country in 1999.[313] The last evaluation well drilled in Eritrea, Chita-1,[314] was by Anglo-French Perenco in 2005-06 and ended in a gas show, according to analysis by a Japanese geologist.[315]

Ethiopia.

Ethiopia is not currently producing oil and gas.[316] The country does, however, have five sedimentary basins in Ethiopia with prospects for hydrocarbons: the Ogaden, Gambella (these first two are the most promising), Abbay, Omo Valley, and Mekele.

Ogaden.

The Ogaden, in the southeast bordering Somalia, has historically been destabilized as predominantly ethnic Somali rebels with the Ogaden National Liberation Front (ONLF) has stated that it would not allow the resources of the region to be exploited. In 2007, the ONLF destroyed an oil exploration facility outside Abole operated by China's Sinopec, killing 65 Ethiopians and nine Chinese. Concerned by the security situation, Sweden's Lundin Petroleum sold its Ogaden assets to Canada's African Oil Corporation in 2009,[317] and Malaysia's Petronas sold out its shares to China's Hong Kong-based PetroTrans in 2010,[318] whose facilities were also threatened with attack from an Ogaden-based rebel movement.[319] The Ethiopian government cancelled five production sharing agreements for 10 blocks with PetroTrans in July 2012, claiming PetroTrans had failed to raise promised financing for new investments.[320]

Despite instability in the Ogaden region, the Ethiopian government had awarded exploration rights to 17 of 21 blocks in the Ogaden Basin as of 2011.[321] African Oil has one project as 30 percent operator in the Ogaden Basin (Blocks 7 and 8), which it describes as "relatively unexplored with limited well and seismic data to constrain a petroleum system proved by the

Calub and Hilala gas condensate fields,"[322] The fields were discovered by Tenneco in 1973 and 1974, respectively,[323] but this U.S. firm was forced to withdraw by the 1974 Ethiopian revolution.[324] African Oil's partners in Blocks 7 and 8 and Adigala are New Age (African Global Energy), with a 40 percent share and the UK's Afren with 30 percent,[325] with the Ethiopian government retaining a 10 percent back-in right into successful exploration. Afren and its partners relinquished Blocks 2 and 6 in order to focus future exploration on Blocks 7 and 8.[326] SouthWest Energy, which describes itself as an indigenous Ethiopian energy company, has won three concessions (Blocks 9, 9A, and 13) to explore for oil and gas in the "Jijiga" (Ogaden) Basin. The firm has sought a $100 million private placement of capital[327] in order to fund drilling scheduled to begin in 2013 2Q.[328] SouthWest, using gravity, aeromagnetic, and seismic data, has concluded that the Ogaden has a newly identified rift system that is Permo-Triassic and Karoo age and part of:

> a prolific petroleum system in East Africa with billions of barrels of oil discovered in Madagascar, several trillion cubic feet . . . of gas condensate in the Ogaden Basin, and a large number of hydrocarbon shows all along the East African coastal basins.

The firm believes its three blocks have two structural traps with potential recoverable resources of over 500 million barrels of oil equivalent (BOE).[329] It believes the Basin as a whole could contain between 1.5 and 6 billion barrels of oil;[330] a "competent persons report" evaluating SouthWest Energy's blocks concluded that the Gambella and Jijiga Basins have best estimate prospective resources of 1.56 billion barrels.[331] Another source estimated that Ogaden could contain

5 tcf of gas deposits, which would give Ethiopia larger reserves than established gas producers such as Cameroon and Congo (Brazzaville).[332] The Calub and Hillala fields alone are estimated to contain 3 tcf.[333]

Gambella.

The Gambella Basin, in the southwest bordering South Sudan, is believed to be an extension of the prolific Melut Basin located in southern Sudan,[334] where hydrocarbon reserves in excess of 600 million barrels have been discovered. The Gambella Basin is part of the Central African rift system that trend across Central Africa from the Benue Trough in Nigeria, through Chad, into South Sudan. In 2008, Canada's Calvalley obtained a 100 percent operator license for the Gimbi Block in the northeastern flank of the Gambella Basin.[335] In January 2012, SouthWest Energy signed a production sharing agreement for the Gambella Block, which at 17,000 km² is half the size of Belgium.[336]

Abbay.

In west-central Ethiopia, with South Sudan to the west, and the capital of Ethiopia, Addis Ababa at the southeast limit of the basin.

Omo.

Anglo-Irish Tullow found indications of hydrocarbons in the first of two exploratory wells drilled 2013 1Q on the South Omo Basin (operator, 50 percent share, with partners Africa Oil, 30 percent, Marathon, 20 percent),[337] which is located in southwest Ethiopia, east of South Sudan and north of Kenya, bordering on

Lake Turkana. This was the first evidence of a working petroleum system in previously undrilled South Omo.[338] Geologically, the Omo Basin is in the northern portion of the Tertiary East African Rift Trend, where Tullow has made two important discoveries in the Lokichar Basin of Kenya.[339] Tullow's six adjacent Kenya licenses, along with the Ethiopia license that it shares with African Oil Corp (30 percent) and the U.S. firm Marathon (20 percent),[340] cover the Turkana Rift Basin, which is similar in geology to the Lake Albert Rift Basin and also a southeast extension of the geologically older Sudan Rift Basin trend.[341] African Oil also has a 50 percent concession in the Rift Basin area north of the South Omo Block, which is an extension of the Tertiary-age East Africa Rift Trend in Ethiopia including South Omo and the Kenyan Blocks.[342]

Mekele.

Mekele is located south of Eritrea. New Age (operator, 40 percent), Africa Oil (30 percent), and Afren (30 percent) share a concession in Adigala close to the borders with Djibouti and Somalia and hope to drill in the El Kuran Prospect in 2013 2Q.[343] Regional geophysical data suggests a sedimentary basin on-trend with the Jurassic Basins of Somalia.[344] In 2008, Calvalley obtained 100 percent operator rights to the Metema Block in the center of the Mekele Basin, which is also part of the Central African Rift System.[345]

Not clearly identified as being in any of these basins is an 80 percent operator concession that Canada's Epsilon Energy won in 2009 for a 82,500 km^2 trade of land in northwest Ethiopia[346] for which it had carried out a high-resolution airborne gravity and magnetic survey as of 2011.[347]

Shale Oil.

Ethiopia has an estimated 3.98 billion tons of oil shale (enough to produce about one trillion barrels of oil) in Tigray State, on the border with Eritrea to the north, and Sudan to the west, and 120 million tons of oil shale in the Delbi.[348] As the result of a joint study conducted by SouthWest Energy and the Ethiopian government, the two negotiated a Production Sharing Agreement (PSA) to begin further exploration on a 13,000 km² block in the Jimma area, which has great potential for the development of oil shale operations.[349]

Gabon.

Oil output in Gabon dropped about 40 percent from its peak of 370,000 bpd in 1997 to 2012, causing Gabon to fall to the third largest oil producer in Sub-Saharan Africa to the sixth largest, Nigeria, Angola, Sudan/South Sudan, Equatorial Guinea, and Congo (Brazzaville).[350] In the short term, declines in production from larger fields will continue to be mitigated by smaller fields; nevertheless, in the long run, Gabon's oil production will depend on the success of new exploration, particularly of deepwater, pre-salt fields, interest in which has been piqued by recent offshore Brazil pre-salt discoveries.[351] Specifically, there has been a conjugation of Brazil pre-salt fields such as the Campos Basin with West African pre-salt formations in Gabon, Kwanza in lower Congo, and Cabinda in Angola.[352]

Energy Minister Etienne Ngoubou announced in December 2012 that Gabon's oil production was averaging 225,000 bpd, and was expecting output to increase slightly in 2013 and 2014. Gabon had proven oil reserves of 3.68 billion barrels at the end of 2011[353] and proven gas reserves of 1 tcf.[354] The country's natural gas resources have not been exploited because Gabon lacks the infrastructure to utilize the natural gas for domestic industry or electricity generation. Nearly all the associated natural gas produced is vented and flared or re-injected. The country produces 73 billion cubic feet (bcf) of natural gas, and only 3 bcf is consumed locally, mainly in two power plants.[355] There are no LNG terminals or projects in the country.[356]

The Anglo-French company Perenco has current production totaling 65,000 bpd through several onshore and offshore licenses. In 2011, the company drilled the first of six new wells in the Oba Field and three exploratory wells in the Arouwe offshore block and the Fernan Vaz Lagoon.[357] Anglo-Dutch Shell produced 65,000 bpd from five fields.[358] It also reached a deal in July 2012 to farm out a 25 percent share of offshore Blocks BC9 and BCD10 to China's CNOOC.[359] China's Sinopec continued work on "several exploration wells" in Block G4-188.[360] Sinopec's Addax subsidiary had five production sharing agreements, two as onshore operator of Maghena and Panthere NZE, with five producing fields, two discoveries, and one under development.[361] Addax is enmeshed in a legal dispute soon to be in arbitration with Gabon over the government's seizure of the Obangue oilfield in January 2013 over a dispute of more than $1 billion concerning customs duties and compliance with other laws.[362]

France's Total was operator of 26 fields and nonoperator of 1 field, cumulatively producing 58,000 BOE in 2011. It carried out a redevelopment project on the Anguille Field in which 21 development wells are to be drilled. Total was the 63.75 percent operator of the deep-offshore Diaba license and is currently analyzing 3D data on the field. In 2010, Total farmed into the onshore Mutamba-Itoru exploration license (50 percent) with the U.S. firm VAALCO energy and with Perenco for DE7 (30 percent) and Nziembou (20 percent), but relinquished the DE7 after negative exploratory drilling. Total is analyzing the other two fields and is carrying out seismic work on the Nziembou license area; Total expects to drill an exploration well on Mutamba-Itoru in 2012.[363] Total also reached a farm out agreement with U.S. firm Marathon, which purchased a 21.25 percent working interest in the offshore Diaba license G4-223.[364] VAALCO had a 28.1 percent interests in exploration acreage in the Etame Marine Block and development areas surround the Etame, Avouma, and South Tchibala oilfields.[365] Vaalco successfully completed a new development well in the Avouma Field in April 2013, with a natural flow rate of 3,000 bpd, and will use the same rig to drill an exploratory well in Prospect Mu north of the Etame Field.[366]

Other foreign oil companies active in Gabon include NOC Oil India Ltd., which has a 45 percent operator interest in onshore interior basin block Shakti (FT-2000) with Singapore Oil (45 percent) and Singapore's Marvis (10 percent).[367] Oil India, which began drilling in Shakti in November 2012, continued talks in February 2013 to acquire France's Maurel et Prom assets in Gabon, which include 85-100 percent interests in five production permits and four exploration permits.[368] Oil India has earmarked about 70 billion rupees ($1.28 billion) for overseas acquisitions to help the nation secure its energy supplies, but has been stymied by lengthy processes for the necessary government approvals.[369] In 2011, Brazil's Petrobras entered the country through an agreement to buy a 50 percent stake from the UK's Ophir Energy in the Ntsina Marin and Mbeli Marin Blocks, in the Coastal Basin, offshore northern Gabon.[370]

In November 2012, U.S. firm VAALCO made an oil discovery while drilling in the onshore Mutambo Iroru Concession. VAALCO has a 50 percent operator interest, while partner Total holds the remaining 50 percent, with the concession subject to an 18 percent Gabon government back-in interest in the event of development.[371] In January 2013, U.S. firm Harvest Natural Resources struck oil at the pre-salt Gamba and Dentale formations offshore in the Dussafu Marin Block, [372] where it has a 66.67 percent operator interest.[373] This most recent discovery, along with three earlier ones in the block at Ruche, Walt Whitman, and Moubenga, brings the combined mean estimate for contingent resources in the block to 45 million barrels.[374] The U.S. firm Cobalt International is pushing its own pre-salt program, partnering with Total, which drilled Gabon's first one deepwater pre-salt well in the Diaba Block in early 2013,[375] (Cobalt also plans for four pre-salt wells off Angola.)[376]

As part of a bid to attract more foreign investment and spur further increases in production, the Minister also announced new, investor-friendly regulations and a new deep offshore round of licensing in June 2013[377] involving new offshore 42 blocks.[378] In January 2013, Minister Ngoubou announced the awarding of an 80 percent interest in the offshore Nkembe Block to Australia's Pura Vida.[379] In March 2013, Pura Vida announced an oil discovery at the Loba M 1 well in Nkembe.[380]

Gambia.

In 2010, Australia's African Petroleum entered into a farm-in agreement to acquire a 60 percent operator interest in two Gambian deepwater offshore exploration licenses, Block A1 and A4. The remaining 40 percent equity is held by the UK/Cyprus firm Buried Hill Energy. There are potential mean unrisked recoverable prospective undiscovered resources of approximately 500 million barrels. African Petroleum plans to drill the first well in the Gambia in 2013 2Q.[381] There are no known oil or gas reserves in the Gambia, but the country has good prospectivity for hydrocarbons, as the area marks the northern extent of the Casamance-Bissau sub-basin, which forms part of the Mauritania-Senegal-Gambia-Guinea Bissau Coast Basin,[382] also known in the industry as the "MSGBC" Basin, which itself is part of the North-West Africa Atlantic Margin. In this regard, Norway's Bergen Oilfield Services won a contract in 2011 with the governments of Senegal, the Gambia, Guinea-Bissau, and Guinea to carry out a 15,000-km, 2D seismic survey from Senegal to Guinea.[383] Oil discoveries in Ghana and later in Cote d'Ivoire and Sierra Leone have encouraged increased, early-stage exploration from Ghana to Guinea-Bissau, and the Gambia appears to have direct analogies to many of the regional exploration successes in West Africa.[384]

In January 2012, the U.S. firm CAMAC energy won two offshore blocks in 600-1,000 meters of water, A2 and A5, which are east of African Petroleum's Blocks A1 and A4. In 1979, Chevron drilled the Jammah-1 well on the basis of sparse 2D data and had gas shows.[385] CAMAC Energy will be the operator with a 100 percent interest, with an option by the Gambia

National Petroleum Company to participate up to 15 percent following approval of a development plan.[386] CAMAC believes that:

> A fresh interpretation of . . . West Africa's petroleum geology . . . raise[s] the prospect of a new oil province in the West Africa Transform Margin, an area between the two tectonic plates stretching from eastern Ghana, Cote d'Ivoire, Liberia and Sierra Leone,

with Gambia having similar geology.[387]

Ghana.

Ghana produced only 8,880 bpd in 2010, but became Africa's newest oil exporter in December of that year when the new Jubilee Field began shipping oil.[388] Ghana's oil production increased almost 10-fold in 2011, reaching 84,737 bpd by year-end. The partners were able to increase production at Jubilee to 105,000 bpd in December 2012 after remedial work and the start-up of the Jubilee Phase 1A development plan,[389] including additional acid stimulation operations on some of the Phase 1 wells. All five Phase 1 production wells were expected to be online by 2013 1Q.[390] The U.S. firm Kosmos expects Jubilee to produce an average of 105,000 to 115,000 bpd in 2013,[391] and later to climb to 200,000 bpd. A Ghanaian government official said in February 2012 that Ghana hoped to boost crude oil output more than five-fold over the next 5 years, to about 500,000 bpd.[392]

Estimates of Ghana's oil reserves vary widely. A Ghana NOC official said in 2011 that Ghana's oil reserves were at least 1.25 billion barrels.[393] Jubilee alone is estimated to contain anywhere from 800 million to 1.8 billion barrels of oil,[394] although Kosmos estimated the field at 600 million BOE to 1 billion barrels.[395]

Ghana has five sedimentary basins: the inland Voltaian Basin, the offshore Accra-Keta Basin with an onshore extension, the offshore Saltpond (Central) Basin, the offshore Cote d'Ivoire-Tano Basin with an onshore extension, and the offshore Cape Three Points Basin (sometimes considered part of the Tano Basin.)[396]

The government of Ghana has published a list of petroleum agreements executed between Ghana National Petroleum Corporation (GNPC) and various companies for petroleum exploration and production rights between 2002 and 2008 for 12 acreages awarded to 10 operator companies:[397]

1. U.S. firm Kosmos Energy, over the Deepwater West Cape Three Points Block. Kosmos (30.87 percent operator) made discoveries at Mahogany (2008), Kosmos made discoveries at Mahogany (2008), Teak (2011), Akasa (2011), and had a technical success at Banda (2012).[398] Other partners are Anadarko (30.87 percent), Anglo-Irish Tullow (22.9 percent), GNPC (10 percent), the EO Group (3.5 percent), and Sabre Oil (1.86 percent).[399] The consortium is in the pre-development plan phase for this cluster of fields known as MTAB (Mahogany, Teak, Akas, and Banda).

2. Tullow, over the Shallow Water Tano Fields, including the discovered North and South Tano Fields; and the West Tano Heavy Oil discovery.

3. Tullow, 49.95 percent operator, over the Deepwater Tano Basin, with partners Kosmos (18 percent), Anadarko (18 percent), Sabre (4.05 percent), and GNPC (10 percent). The world-class Jubilee Field was discovered in 2007 and reached first production in 2010 — a record time. The field straddles both the deepwater Tano and West Cape Three Points Blocks.[400] Tullow's share is 34.71 percent, Anadarko and Kosmos each have 23.49 percent, GNPC has 13.75 percent, Sabre Oil

& Gas Holdings has 2.81 percent, and and E. O. Group has 1.75 percent.[401]

The Ghanaian government approved in May 2013 a development plan for a new cluster of fields in deepwater Tano known collectively as TEN (Tweneboa, discovered in 2009), Enyenra (2010), and Ntomme (2011).[402] The capital expenditures for the TEN project are around $4.5 billion, which includes around 23 injection and production wells and excludes floating production, storage, and offloading (FPSO) lease costs—a figure so large that Tullow hopes to sell a piece of its interests in the project to raise financing. First production is expected within 36 months[403] with peak production of 80,000 bpd.[404] Kosmos also made a discovery at the Wawa-1 exploration field in deepwater Tano in August 2012, which may lead the firm eventually to co-develop a fourth cluster of fields in Ghana, along with Jubilee, MTAB, and TEN.[405]

4. The Netherlands' Vitol, over the Offshore Cape Three Points area, east of the Kosmos acreage. ENI took over as operator (47.22 percent interest), with Vitol (37.78 percent) and GNPC (15 percent, with an option for an additional 5 percent).[406] ENI discovered gas at both Sankofe East 1-X[407] (in September 2012) and Gye Nyame 1 exploration wells, and is interested in fast-tracking the development of Sankofe,[408] including by drilling other wells to delineate the size of the discovery and confirm commercial feasibility.[409]

5. Vitol, over the offshore Cape Three Points South. In 2009, Vitol made a discovery at Sankofa-1A.[410] ENI took over as operator (47.22 percent interest), with Vitol (37.78 percent) and GNPC (15 percent).[411]

6. U.S. firm Hess, over Deepwater Tano/Cape Three Points (south of the Tullow and Kosmos Blocks), is a 55 percent operator, with Norway's Statoil holding 35 percent interest and GNPC holding 10 percent.[412]

In 2011, oil and gas condensates were found at the Paradise-1 exploration well,[413] with later discoveries at Almond (oil), Beech (oil), Hickory North (gas condensate), and, in December 2012, at Pecan-1 (oil).[414] Based on the results of these wells and Hess' experience in Equatorial Guinea, where the geology is similar, the firm plans to submit appraisal plans to the government by June 2013.[415]

7. U.S. firm PanAtlantic (formerly Vanco), over the Cape Three Points Deepwater Block. Russia's Lukoil farmed into a 56.66 percent interest, with PanAtlantic retaining a 28.34 percent interest, and GNPC with a 15 percent carried interest and the option to acquire up to a further 5 percent in any commercial discovery. In 2010, PanAtlantic made an oil and gas-condensate discovery at the Dzata-1 exploratory well.[416] Lukoil subsequently drilled two wells with noncommercial hydrocarbon reserves and plans to drill another well in 2013 4Q,[417] with the hope of cumulatively finding 150-200 million barrels of oil—enough to make the production profitable.[418]

8. Norway's Aker ASA, 85 percent operator of the Deepwater Tano South contract Area, which is the western part of the area relinquished by PanAtlantic. Partners are GNPC (10 percent) and Chemu Power, a local Ghanaian Company (5 percent). South Deepwater Tano is located further offshore in the Gulf of Guinea and at greater water depths than Tullow's Jubilee field.[419] PanAtlantic planned one exploration and 1-3 appraisal wells in 2012-2013.[420]

9. India's Gasop Oil, over Deepwater Saltpond Basin. Gasop is owned by OMEL, which in turn is a joint venture between India's ONGC and Mittal.[421]

10. Saltpond Offshore Producing Company Limited, over the existing Saltpond Field. This company is

a joint venture between the U.S. firm Lushann International, Nigeria's Eternit, and GNPC. Lushann-Eternit Energy Ltd. has a 55 percent share, while GNPC has 45 percent. This company is Ghana's oldest oil production company and the Saltpond Field has been in production intermittently for over 25 years. Daily production is currently about 600 bpd,[422] down from a peak of 4,800 bpd. In 2000, GNPC had entered into an agreement with Lushann International for the rehabilitation of the Saltpond Oil Field.[423]

11. Nigeria's Oranto Petroleum International Limited, over the Saltpond area.

12. Italy's ENI, over the deepwater Keta Basin, with a 35 percent operator interest, with the UK's Afren (35 percent), Japan's Mitsui (20 percent), and GNPC (10 percent). ENI took over from Afren as operator in 2011. The Keta Block is in the offshore part of the Volta River Basin in eastern Ghana, next to the maritime boundary with Togo.[424] The consortium failed to encounter hydrocarbons in Nunya-1X exploration well,[425] but believes the basin retains a high potential and plans to drill another exploratory well in 2014.[426]

Other Fields.

Two other exploration areas are offshore Accra and the onshore Voltaian (or Volta River) Basin. Australia's Tap Oil has a 45 percent interest in offshore Accra acreage.[427] In December 2012, the UK's Ophir Energy (new operator), Rialto (12.5 percent share), and Vitol farmed-in to this acreage, which shows similarities to Jubilee, and planned to drill the Starfish-1 exploration well in 2013.[428] Tap said the well had a P50 (50% probability) resource of 431 billion barrels and could be comparable in size to Jubilee.[429] Canada's Simba Oil has submitted an application for acreage in east-central Ghana in the Voltaian Basin and hopes it could "represent a large gas play."[430]

National Gas Processing Project.

In July 2012, China's Sinopec won a $750 million contract to develop Ghana's National Gas Processing Project, which involves transporting and marketing associated gas from the Jubilee Field. The Chinese government offered financing and agreed to accept repayment in the form of a long-term oil supply contract. Chinese firms are involved in several other energy-related projects, including thermal power plants and the Bui hydroelectric project.[431] Gas will be distributed from a central processing facility at Domunli; a 120-km onshore gas line will deliver dry gas to the Aboadze thermal power station, while another 75-km onshore line will connect to the mining center of Prestea.[432] Sinopec, for example, is the contractor for a project to be completed by July 2013 to build a pipeline from the offshore Jubilee Field to deliver 150 million cf/d to the Aboadze plant.[433]

Territorial Dispute with Cote d'Ivoire:
No Impact on Exploration and Production.

Ghana has a long-standing maritime territorial dispute with Cote d'Ivoire that reignited in 2010 after Ghana started commercial oil production. Cote d'Ivoire has been claiming ownership of billions of barrels of oil and cubic feet of gas reserves found in deep waters near the coast of Ghana and covering portions of the Jubilee, Field, Tweneboa, Enyenra, and Owo discoveries and West Tano-1X find.[434] Both countries had previously made submissions in 2009 to the UN Commission on the Limits of the Continental

Shelf.[435] IOCs continue work in the disputed areas under the assumption that they belong to Ghana, given that the UN's International Court of Justice has not issued an injunction against Ghana pending its adjudication of the dispute.

Guinea.

Guinea does not produce oil or gas, but there is ongoing offshore and onshore exploration.[436] The U.S. firm Hyperdynamics holds one of the largest offshore exploration acreage positions in West Africa, at 25,000 km², with depths from 100 meters to 4,000 meters.[437] It has acquired 17,800 km of 2D seismic and 3,635 km of 3D seismic data.[438]

The company relinquished a portion of its original, 31,000-square-mile concession in 2009 under the then military government.[439] According to the U.S. Foreign Commercial Service, Hyperdynamics relinquished 70 percent of its original concession to the Transitional government. Some speculated that officials in the Transitional government had awarded much of this offshore acreage to China Sonagol, a joint venture between the Angolan NOC and Duyuan International Development. However, the government of Alpha Conde, who, in December 2010, became the first democratically elected President in Guinea's 50-year history, said it will review agreements made by previous regimes.[440] China Sonangol's website does not list any oil and gas operations in Guinea, but does list a Conakry office.[441] In January 2010, Hyperdynamics farmed out 23 percent of its concession to Dana Petroleum, a wholly owned subsidiary of the Korea National Oil Company; in February 2013, it completed the farm out of 40 percent to Anglo-Irish Tullow Oil. CEO Leonard said that:

We look forward to working with Tullow Oil as they apply their successful exploration experience on the Atlantic Margin . . . particularly the Transform Margin play that is present in the Guinea acreage.[442]

The partners intend to drill a deepwater well before April 2014.[443] Earlier, in February 2012, Hyperdynamics announced that its exploratory well Sabu-1 had encountered hydrocarbons that its CEO Ray Leonard indicated "enhances the prospectivity of our . . . concession."[444]

In 2011, Canada's Simba Energy signed an agreement with Luxembourg-based Summa Energy for a 60 percent operator share in 12,000 km^2 in onshore Blocks 1 and 2 in Guinea's Bove Basin,[445] which has three known reservoir systems.[446] During the firm's initial exploration work in February 2013, Simba identified numerous oil seeps in both blocks.[447] It plans to drill the first of two test wells later in 2013.[448]

The government of Guinea announced that it plans to put 22 new offshore blocks up for auction after it puts its new petroleum code in place, hopefully by June 2013.[449]

Guinea-Bissau.

While there is currently no oil or gas production in Guinea-Bissau, there have been oil discoveries in both the country's offshore blocks and its Joint Development Zone (JDZ) with Senegal. The Zone, principally known by its French name, *Agence de Gestion et de Cooperation entre la Guinee-Bissau et le Senegal (AGC)*, was established under the terms of a 1995 agreement that addressed a territorial dispute between the two countries and divided proceeds from activity in the joint exploration area between Guinea-Bissau and Senegal in a 15:85 ratio.[450]

There has been intermittent exploration offshore in Guinea-Bissau since the 1960s, when Esso drilled six wells. In 1974, Guinea-Bissau gained its independence from Portugal. Since then, exploration has frequently been affected by civil unrest. Offshore exploration had also been hampered by a boundary dispute with Senegal, which was not resolved until 1993.[451] In recent years, perceptions of Guinea-Bissau's reputation as a target for foreign investment has been severely damaged by perceptions that it had devolved into a "narco-state" in which Latin American drug cartels had corrupted the weak central government and was using the country as a transit point for cocaine shipments from Latin American to Europe.[452]

Despite this, there are at least three groups of foreign oil companies engaged in exploration in Guinea-Bissau and the AGC. The first group is led by Svenska Petroleum, a Swedish company that is 100 percent owned by Saudi Arabian/Ethiopian businessman Sheikh Mohammed H. al-Amoudi.[453] Svenska is the operator of the Sinapa (Block 2) and Esperanca (Blocks 4A and 5A), along with Australian partner FAR (21.43 percent) and NOC Petroguin.[454] These licenses lie on the continental shelf around 180 km off the Guinea-Bissau coast and west of the Bissau River estuary. Immediately to the north lies the billion-barrel Dome Flore discovery. The blocks have wide, shallow waters of an extended continental shelf, where the Sinapa oil discovery, made in 30 meters (m) of water, is estimated to have P50 STOOIP of 240 million barrels; several large untested prospects in the Sardinha Prospect are estimated to have unrisked P50 STOOIP of 219 million barrels. Svenska plans to drill two wells in 2013.[455] Qatar's Sphere Petroleum, a now privately held company formerly traded on the Australian Stock Exchange, is the operator of offshore Blocks 1 and 5B.[456]

AGC: The UK's Ophir is the lead operator (44.2 percent) in deepwater Block AGC Profond, along with partners U.S. firm Noble (30 percent), L'Entreprise (AGC state entity; 12 percent),[457] FAR (8.8 percent), and the U.S. firm Rocksource Energy (5 percent).[458] Ophir drilled an unsuccessful well in June 2011 and has undertaken analysis of the wider implications for the prospectivity of the Casamance Sub-basin of the MSGBC Basin.[459]

Kenya.

Major oil discoveries in Uganda in 2006 and natural gas discoveries in Mozambique and Tanzania in 2010 have created expectations that Kenya is part of a new regional hydrocarbon province, attracting new IOC exploration. Kenya has four large sedimentary basins:

- Anza, in north central Kenya, with its northern tip touching Ethiopia and running northwest-southeast, touching Somalia.[460]
- Lamu, in onshore and offshore in southeast Kenya, may contain up to 5 billion barrels of oil. Geology similar to areas of Madagascar, the island that was joined to East Africa before splitting apart 145 million years ago and that may contain up to 24 billion barrels of heavy oil resources.[461] At least 11 wells in Lamu have encountered oil and/or gas shows.[462]
- Mandera in the northeast tip of Kenya, bordering Ethiopia to the north and Somalia to the east.

- The Tertiary Rift Basin in western Kenya, bordering Sudan and Ethiopia to the north, Uganda to the east, and Tanzania to the south.[463] Includes several sub-basins, including Lochikar, where Anglo-Irish Tullow Oil made two oil discoveries in 2012.[464] Kenya's Great Rift Valley could yield 10 billion barrels of oil, Tullow estimates.[465]

Until recently, only 28 exploration wells had been drilled in Kenya, averaging one well every 20,000 km², with the earliest being drilled in 1992. The rush is now on, however, and there are already at least 12 oil exploration companies in Kenya, including France's Total, Italy's ENI, and Norway's Statoil.[466] As of July 2012, the government had distributed 46 designated blocks, including nine offshore.[467] In February 2013, the government announced that it would adopt an auction-style format for up to nine new onshore and offshore blocks. It had previously operated more on a first-come-first-served basis, but is demanding higher licensing fees and work program requirements now that Kenya has become an established area for hydrocarbons. Kenya's production sharing contracts say that explorers must relinquish 25 percent of the acreage in a licensed area every 2 years. Anglo-Irish Tullow Oil and Anadarko reached this point in 2012 for seven exploration blocks, which the government now plans to re-demarcate into up to nine new blocks.[468] The government has a $25 billion plan to construct a new port in the northern coastal town of Lamu that could be used for energy and other exports from Kenya, Uganda, Ethiopia, and South Sudan.[469]

Onshore.

Tullow Oil and Africa Oil (50 percent each) found crude at two onshore wells in March 2012 in the northern Turkana region in Blocks 10 BB (Ngamia-1 exploratory well) and 13T (Twiga South-1),[470] and now plan up to 11 more test wells in 2013. The IMF has already declared these Turkana finds as "commercial" and projects production of oil will start in "6-7 years," while Tullow has been more cautious[471] — perhaps in part to tamp down Kenyan government expectations and desire to accrue more benefits to itself through new legislation. Tullow described this area as in the South Lokichar Basin, one of seven sub-basins in the larger Kenya-Ethiopia Rift Basin. Each of the seven is similar in magnitude to the Lake Albert Rift Basin in Uganda. Tullow also has 50 percent operator interests in Blocks 10A and 10BA, a 65 percent operator interest in Block 12A, and an undisclosed farmed-in, working interest in Block 12B,[472] where Australia's Swala Energy has carried out an airborne gravity-magnetic survey and plans a 2D survey soon.[473] In March 2013, Tullow found light, noncommercial hydrocarbons in Block 10A in the Anza Basin in northeast Kenya, with Canada's Africa Oil (30 percent) and the UK's Afren (20 percent).[474] Canada's Africa Oil also owned shares running from 20 percent to 50 percent of Blocks 9, 10BA, 10BB, 12A, and 13T, and agreed in September 2012 to sell to U.S. firm Marathon Oil a 50 percent interest in Block 9 and a 15 percent interest in Block 12A.[475] As with Block 10A, the firm's Block 9 is in the Anza Basin, along a northeast-southwest trend rift that runs from southern Sudan.[476] These two blocks had prospective gross best estimates of 1.287 billion and 588 million barrels of oil, respectively.[477] Afren has an 80 percent operator interest in Block 1 in northeast Kenya along the border with Ethiopia and Somalia,[478] along with Canada's Taipan Resources, which also has a 100 percent interest in Block 2B, to the south bordering with Somalia. The Tarbaj oil seep in southwest Block 1 suggests an active petroleum system, while the northern portion is an extension of the Ogaden Basin in Ethiopia, which is estimated to contain approximately 4 tcf of natural gas and 715 million BOE. Block 2B is estimated to have 387 million BOE of unrisked prospective resources.[479] Taipan expects to drill its first well on Block 2B in earlier 2014.[480] Canada's Imara Energy acquired Block L2 onshore Lamu Basin in June 2012.[481] Privately held Swiss Oil Holdings International owns the licenses to Blocks L4 and L13, which are both mainly onshore at Lamu (L4) and further southwest along the coast.[482]

In July 2012, U.S. firm ERHC was awarded Block 11A in northwestern Kenya along the border with South Sudan, east of where Tullow made discoveries in 10BB.[483] Block 11A has parts of both the Central Africa Rift System and the Tertiary East Africa Rift System; the firm says that the proximity and intrend relationship between the Abu Gabra Rift Basin of southern Sudan and the Lotikipi Sub-basin in the East Africa Rift System suggest high oil and gas prospectivity.[484] In May 2013, ERHC announced a letter of intent to farm out part of the block to an unnamed "integrated international oil and gas company" while it moves forward on a full tensor gravity gradiometry survey of the block.[485]

Canada's Simba Oil has a 100 percent share of Block 2A; the southwest corner of the block extends into the Anza Basin in central Kenya, moving northeasterly toward the southern tip of the Mandera Basin in northeastern Kenya bordering on Somalia.[486] Simba expects to strike oil reserves of over one billion barrels in Wajir Region.[487] In May 2013, Simba farmed out a 66 percent share and operatorship of the block to privately owned oil and gas company, APEX Exploration.[488] Canada's Vanoil has Blocks 3A and 3B due south of Block 2A, bordering with Somalia to the east. These blocks are at the confluence of the Anza, Lamu, and Mochesa Basins.[489] The firm also later acquired shares in Blocks 9, 10A, 10 BA, and 10BB, the latter two containing most of Lake Turkana.[490] The UK's Bowleven farmed into a 50 percent interest in Block 11B, which borders on South Sudan and Ethiopia and is west of Lake Turkana.[491] In 2010, CNOOC plugged the Bhogal-1 exploration well on Block 9 in north-central Kenya after inconclusive results showed hydrocarbon potential.[492]

Offshore.

In September 2012, U.S. firm and 50 percent operator Apache found a major offshore discovery of gas in offshore Block L8 about 70 km east of Malindi at the Mbawa-1 exploration well,[493] with Australia's Origin Energy (20 percent), Australia's Pancontinental (15 percent), and Tullow Oil (15 percent).[494] Pancontinental says Mbawa has "world-class potential for oil and gas, with volumetric pressure easily exceeding one billion barrels of oil" and potentially exceeding 5.2 billion barrels and 809 bcf of gas."[495] U.S. firm Anadarko is the 50 percent operator of Blocks L5, L7, L11A, L11B, and L12, with partners Total of France (40 percent) and PTTEP of Thailand (10 percent).[496] In May 2012, U.S. firm CAMAC Energy was named 100 percent operator of Blocks L1B, L16, L27, and L28, with the Kenyan government having the option to participate to

20 percent upon development,[497] and began airborne gravity and magnetic surveys of Blocks L1B and L16 in April 2013.[498] Australia's FAR has a 60 percent operator interest in Lamu Basin offshore Block L6, where 3D data has led to an unrisked estimate assessment of as much as 3.96 billion barrels of oil or 10.69 tcf of gas, with drilling planned for mid-2013.[499] FAR, which recently convinced the Kenya government to restore a large area in Block L-6 that had been previously relinquished, estimates that L-6 has prospective resources of 3.96 billion barrels of oil or 10. 7 bcf of natural gas.[500] FAR also holds a 30 percent interest in offshore Block L9,[501] which is operated by the UK's Ophir Energy. Ophir plans to drill a well in Block L9 in 2013.[502] In the southern, offshore portion of the Lamu Basin, the UK's BG Group is the 40 percent operator of Block L10A (with partners Thailand's PTTEP, 25 percent; the UK's Premier, 20 percent; and Pancontinental, 15 percent) and L10B (Premier, 25 percent; PTTEP, 15 percent; and Pancontinental, 15 percent).[503] BG Group will drill at least one well in L10A/L10B in the second half of 2013.[504] Afren has 100 percent interest in L17 and L18, both of which are in the offshore part of the Lamu Basin adjacent to the coast, with L18 further east offshore, and also abutting the maritime boundary with Tanzania,[505] as well as Afren's Tanga Block in northern onshore and offshore Tanzania.[506] The company obtained 3D seismic in 2012 for L17/L18, prior to exploration drilling.[507]

Lesotho.

There are no known hydrocarbons in Lesotho.

Liberia.

Offshore.

There is currently no oil or gas production in Liberia, but considerable optimism about the possibility of a new hydrocarbon province offshore known as the West Africa Transform Margin.[508] In 2004, Spain's Repsol acquired rights to offshore Blocks 16 and 17, the latter of which runs up to the maritime border with Sierra Leone.[509] In 2010, U.S. major Chevron acquired three offshore blocks, LB-11, LB-12, and LB-14, in which it currently has a 45 percent operator interest, with Nigeria's Oranto with 30 percent and Italy's ENI with 25 percent.[510] The U.S. firm Anadarko has 47.5 percent operator shares in Blocks LB-15, LB-16, and LB-17 in fields adjacent to Sierra Leone's Mercury and Venus discoveries. Other partners are Repsol (27.5 percent) and Anglo-Irish Tullow Oil (25 percent).[511] Anadarko drilled the unsuccessful Montserrado-1 well in Block LB-15 in 2011.[512] Despite this disappointment, partner Tullow stressed that it remained optimistic about Liberia (and Sierra Leone) as part of an overall exploration program in West Africa.[513]

Australia's African Petroleum made a new offshore oil discovery in February 2013 when it drilled the Bee Eater-1 well on offshore Block LB-09. The company is finalizing negotiations with CNPC (through its Hong Kong-listed arm, PetroChina) for an investment of up to 20 percent equity in this block.[514] The Bee Eater-1 well indicated that the oil play discovered while drilling Narina-1 in early 2012, located 9.5 km southeast, extends over a very large area.[515] One recent estimate, which may have been revised upwards, is that

LB-09 contains 500 million barrels of oil.[516] African Petroleum also owns a 100 percent share in block LB-08, which is east of LB-09 and northwest of Ghana's Jubilee Field.[517]

In March 2013, ExxonMobil purchased an 80 percent operator share offshore in Block LB-13 from Canadian Overseas Petroleum Ltd., which retained a 20 percent share.[518] One report estimated that the block has 2.32 billion barrels of gross prospective recoverable resources.[519] Norwegian seismic player TGS has commenced a 3D multi-client survey covering up to 7,800 km² of highly prospective acreage in the Harper Basin, offshore Liberia. TGS's data processing will be available to clients before a 2013 4Q bid round by the Liberian government for new exploration blocks, including ultra-deepwater.[520] Liberia currently has 17 offshore exploration blocks and plans to creation 13 more deepwater blocks.[521]

Onshore.

Canada's Simba energy has a 90 percent share of the onshore reconnaissance license NR-001 covering 1,386 km² of onshore coastal strip of Liberia lying with the Roberts and Bassa Basins. In 2010, an exploration team carried out an extensive oil seep survey of most of the license area.[522]

Libya.

In the mid-2000s, IOCs involvement in Libya experienced a resurgence as the United Nations (UN), and the United States and other nations lifted sanctions in 2003 and 2004, respectively.[523] Libya was Africa's second biggest oil producer,[524] after Nigeria, and produced an estimated 1.65 million bpd prior to the movement in February 2011 to oust long-time dictator Muammar Qadhafi.[525] This was about 2 percent of global production.[526] This was well below the three million bpd in the late 1960s, but an improvement over the 1.4 million bpd in 2000. Libya also produced an estimated 140,000 bpd of non-crude liquids, including condensate and natural gas plant liquids. Output exceeded Libya's OPEC target of 1.47 billion bpd. Because Libyan oil and natural gas exports were almost totally disrupted during several months of internal fighting, the International Energy Agency coordinated a release of 60 million barrels of oil from member-state emergency stocks. Libya oil production started its recovery in September 2011, production was estimated to have recovered to 1.4 million bpd by May 2012 and was seen to be on track to reach pre-conflict levels of 1.6 million bpd by April 2013.[527] In 2010, Libya had targeted production of 2.5 million bpd by 2015. Following the disruption in exploration and production, Libya has set a more modest target of 2.0 million bpd over the next 5 years. Libya's domestic oil consumption is about 300,000 bpd.[528]

Libya is the fourth ranked African country in terms of crude oil production after Nigeria, Algeria, and Angola. About two-thirds of oil production — and over 80 percent of reserves[529] — are from fields in the eastern part of the country, in the Sirte Basin, followed by 25 percent in the southwest Murzuq Basin, and most of the remainder from the Pelangian Shelf Basin near Tripoli, with small fields in the western Ghadames Basin. Libya's proven oil reserves of 47.1 billion barrels as of January 2012 were the largest in Africa,[530] seventh ranked globally, and represented 3.4 percent of global reserves.[531] The next five African countries in terms of reserves are Nigeria (37.2 billion), Algeria

(12.2 billion), Angola (9.5 billion), Sudan/South Sudan (5 billion), and Egypt (4.4 billion).[532]

Libya's rank as a producer and reserve holder is less significant for natural gas than it is for oil. Libya produced 1,069 bcf of natural gas in 2010, of which 825 bcf was marketed, 120 bcf was re-injected, and roughly an equal amount was vented or flared. Domestic gas consumption was 242 bcf. Most production is exported, almost entirely via the Greenstream pipeline under the Mediterranean to Italy. Libya's natural gas reserves in January 2012 were 52.8 tcf. IOCs are less active in natural gas, with ENI being an exception because of the large Western Libya Gas Project, which produces 1.0 bcf/d. Prior to 2011, the NOC had announced plans to increase the country's natural gas production, including from the Faregh Field operated by Waha in the Sirte Basin and Lellitah's offshore Bouri Field.

Libyan law requires foreign companies to have a local partner, either state owned or private, that owns a minimum of 35 percent in any joint venture. There are at least 28 IOCs active in Libya.[533] Among the more important joint ventures are:

1. Waha Oil Company: Consortium of U.S. firms ConocoPhillips (16.33 percent), Hess (8.13 percent), and Marathon (16.33 percent), in partnership with Libya's NOC (59.16 percent).[534] Total capacity is 350,000 bpd, all from fields in the Sirte Basin. Three major growth projects under development by the co-venturers include Faregh II, North Gialo, and NC-98. [535] (Hess is also the 100 percent operator of Area 54 offshore in the Sirte Basin, where hydrocarbons were discovered in 2008.)[536]

2. Akakus Oil Operations: Led by Spain's Repsol, with 50 percent ownership by the NOC. Production capacity about 350,000 bpd; located in the Murzuq

Basin. Overall, Repsol has 9 blocks, of which 7 are exploratory and 2 are in operation. In January 2011, Repsol made a discovery in Block NC-115.[537]

3. Mellitah Oil & Gas: Led by Italy's ENI; operations in Sirte, Murzuq, and Ghadames Basins. Near Zwara in western Libya. Production capacity of 300,000 bpd, of which 50,000 is condensate. Was attacked in May 2013 by militiamen, who stole weapons and military vehicles.[538]

4. Wintershall: This German company, with Russia's Gazprom and the NOC. Production capacity of over 100,000 bpd. Wintershall is currently cooperating with the NOC and Libya's AGOCO to build a new 55-km-long oil pipeline to connect the Wintershall concession C96 with the Amal field, from which it will be exported via the town of Ras Lanuf.[539] (Gazprom obtained 49 percent shares in Blocks C96 and C97 from Wintershall. Gazprom also has Block 19 offshore, for which it is preparing drilling, and Block 64, which is 300 km south of Tripoli in Ghadames Basin.)[540]

5. Harouge Oil Operations: Led by Canada's Suncor with a 49 percent interest;[541] capacity of about 100,000 bpd.

6. Mabrouk Oil Operations: Led by France's Total and NOC, with Norway's Statoil. Capacity of about 70,000 bpd. The NC 186 license in the Murzuq area is operated by Total and consists of seven fields (A, B, D, H, I/R, J, and K).[542] (Total also has non-operator shares in production in five offshore and three onshore zones, and submitted a development plan for Block NC 191 in onshore Murzuk Basin (100 percent operator).[543] For example, Total has a 20.25 percent stake in Block C 137, together with Mabruk Oil (73 percent operator) and Wintershall (6.75 percent).[544]

7. Zueitina: U.S. firm Occidental Petroleum and Austria's OMV; in Sirte Basin. OMV acquired 25

percent of Occidental's Libya assets in 1985. In 2008, Occidental, OMV, and Libya NOC signed one of the biggest redevelopment projects worldwide for these mature fields.[545] Capacity of about 60,000 bpd.

8. MedcoEnergi: The Nafusah Oil Operations joint venture between this Indonesian IOC (50 percent operator), NOC, and the Libyan Investment Authority (50 percent) for Area 47 is one of the few significant projects expected to come online over the next 4 years, starting in 2016 and peaking production at 50,000 bpd.[546] Eighteen out of 20 exploration wells discovered oil/gas; gross contingent resources of 351.7 million BOE.[547] Capacity to be from 50-100,000 bpd.

9. Other IOCs in Libya include Oil India, which has a 50 percent operator interest in onshore Blocks 86 and 102 (4) and a 25 percent working interest in onshore Area 95/96, Block 2/1, 2, and 4.[548] India's ONGC Videsh also has a 49 percent stake in exploration Block NC-189 (with 51 percent owned by Turkey's NOC) and Contract Area 43 in the Cyrenaica offshore basin. China's CNPC shares an interest with Libya's NOC in offshore Block 17-4 in the Pelagian Basin off the northwestern coast of Libya, with a water depth of 200-400 meters.[549] Its subsidiary, Greatwall Drilling Company (GWDC), is also present in Libya.[550] U.S. major ExxonMobil returned to Libya in 2005 after a 25-year absence, winning a license to explore an area covering 2.5 million acres in the Cyrenaica Basin, one of the country's largest energy fields.[551]

10. UK/Netherlands major Shell said in November 2012 that it remained interested in oil and gas exploration opportunities in Libya after abandoning drilling in two blocks, LNGDA and area 89, earlier in the year.[552] In 2010, the U.S. firm Chevron also did not extend its five-year oil and gas licenses in Libya.[553] One major

concern for IOCs is the possibility of a terrorist attack on their facilities, with workers kidnapped or killed, such as happened in February 2013 at the Amenas gas facility in Algeria. A more likely security problem may be in-fighting between militia for influence within the government's Petroleum Facilities Guard.[554] Labor unrest and protests are other concerns, as protesters forced the closure of three major oil terminals in July 2012, and the Waha Field was blockaded for almost 3 weeks in March 2013 until the oil minister intervened to address the demands of 100 workers.[555]

Madagascar/France's Juan de Nova Islands.

Madagascar has five basins, the most important of which are the onshore basins of Ambilobe, Majunga, and Morondava along the western coast.[556] The Ile St. Marie Basin on the northern end of the island and the Cap St. Marie Basin on the south coast have potential but unproven hydrocarbon systems.[557] Political instability has been a serious impediment to foreign investment in Madagascar's oil and gas industry and only made worse by threats from the transitional government to expropriate major oil fields.[558] Madagascar has been rocked by political crisis since 2009, when Andry Rajoelina ousted President Marc Ravalomanana with the help of dissident soldiers.[559] Nevertheless, there were 17 oil companies exploring the island as of 2012.[560]

Onshore.

Madagascar's largest oil and gas fields are the Bemolanga and Tsimiroro fields in the Morondava Basin. Rajoelina's transitional government announced in December 2010 that it was interested in acquiring Blocks 3104-3107, where these giant fields are located, from the U.S. firm Madagascar Oil at less than market prices.[561] The resulting uproar among IOCs led the transitional government to delay a planned tender of 225 offshore blocks in March 2011.[562] Later that month, Madagascar Oil declared force majeure to safeguard its rights under its PSCs,[563] eventually forcing the government to back down, when it acknowledged the firm's rights under the PSCs for Block 3104 in June 2011[564] and for Blocks 3105-3107 in April 2012.[565] In September 2011, Madagascar Oil increased its estimates for contingent original-oil-in-place (OOIP) for Tsimiroro to 1.78 billion barrels.[566] The firm has a Steam Flood Pilot project at Tsimiroro[567] under which it will use injected steam into the ground to soften the oil, with production starting in 2013 1Q and peaking in 2013 4Q or 2014 1Q with 1,000 bpd. Full-scale commercial production at the Tsimiroro heavy oil field is expected to commence sometime after 2017 and eventually reach peak production of around 160,000 bpd.[568] While Tsimiroro has potential for light oil and natural gas, Bemolanga is a field with ultra-heavy oil with potential resources of 16.6 billion barrels (about 9.8 billion barrels of recoverable reserves).[569] France's Total has a 60 percent operator interest for an exploration license in Block 3102 (Bemolanga),[570] with partner Madagascar Oil (40 percent). Total now has the right to search for conventional oil as well; it expects first production of heavy oil by 2019.[571]

The UK's Afren has a 90 percent operator interest in Block 1101, which has net prospective resources of 631 million BOE.[572] (Candax owns the remaining 10 percent.)[573] The block is located in the eastern flank of the Ambilobe Basin in northern Madagascar, whose southwestern portion is adjacent to Exxon's Ampasindava Block.[574] Australia's Caravel Energy has a 25 percent interest in the onshore Behaza Oil Project (Block 3114) in the north, with the ability to increase its share to 80 percent. Two leads in the block are estimated to contain between 22.1 million and 2.48 billion barrels.[575] (The 75 percent owner is reportedly Dr. Rsolovoahangy, a Madagascan national.)[576] The UK's Avana holds a 100 percent interest in licenses in the Sambaina shale oil area of onshore Madagascar.[577] Anglo-Irish Tullow Oil two onshore blocks in the southern Morondava Basin,[578] Block 3109 (Mandabe license) and Block 3111 (Berentry license), for which the firm has acquired 2D seismic data and plans for at least one well in 2013.[579] Mauritius-based, London-listed Essar Energy is 100 percent operator of three onshore blocks in the Morondava Basin.[580]

Offshore.

ExxonMobil operates three blocks inside Madagascar, although it reportedly declared force majeure in 2009 due to the political situation in the country following a nondemocratic change in government, but is considering resuming exploration there, including drilling a well in 2013.[581] Exxon has a 70 percent interest in the offshore Ampasindava and Ambilobe Blocks in the northwest (with 30 percent owned by Sterling Energy) and a 50 percent interest in the west-central offshore Majunga Block (with 30 percent owned by BP, 10 percent by Korea's SK Corporation, and 10 percent by NOC PetroVietnam.[582] The large Sifaka Prospect in Ampasindava has been independently estimated to contain gross unrisked best estimate prospective recoverable resources of 1.2 billion barrels.[583] Canada's

Niko Resources has the offshore Grand Prix Block[584] in west-central Madagascar in the Morondava Basin, which is south of Majunga. The onshore part of the basin is a proven petroleum province with onshore discoveries of oil sands and subsurface heavy oil deposits exceeding 20 billion barrels. The offshore portion of the basin is considered to be part of the same petroleum system but extending into a deeper geologic setting that may yield lighter oil discoveries.[585] Nigeria's South Atlantic Petroleum (SAPETRO) has a 90 percent operator interest in the Belo Profond PSC in deepwater Madagascar, in the Mozambique Channel, with U.S. partner MAREX owning the remaining 10 percent.[586]

France's Juan de Nova Islands.

In the Mozambique Channel between Mozambique and Madagascar lies the Juan de Nova islands, which is controlled by France as part of its TAAF or Terres Australes et Antarctique Francaises (French Southern and Antarctic Lands) administrative region. The islands, which are claimed by Madagascar, were not part of Madagascar when the country gained independence in 1960. The Juan de Nova permit, which is 75 percent owned by Nigeria's South Atlantic Petroleum,[587] is located east of the Rovuma Basin in Mozambique, where Anadarko and ENI have made major gas discoveries, and west of Madagascar's Belo Profond permit.[588] To the east of the Juan de Nova permit and west of Belo Profond is the Juan de Nova Est permit, which was issued by France to the UK's Wessex (70 percent operator) and Australia's Global Petroleum (30 percent). France has declared a continental shelf surrounding the islands and claims the energy and mineral rights.[589] The Juan de Nova Est permit lies 100 km northwest of the Tsimiroro and Bemolanga fields.[590]

Malawi.

Malawi currently has no oil or gas production, but has considerable hydrocarbon prospectivity. The government awarded exploration Block 1 in December 2012 to South Africa's SacOil.[591] Block 1 is located in northwestern Malawi, bordering Tanzania to the north and Zambia to the west. The license is located on trend with the East African rift system, which is a proven exploration province with prolific oil discoveries in Sudan, Chad, Kenya and Uganda.[592] Earlier and more controversially, Malawi awarded Blocks 2 and 3 on Lake Malawi to the UK's Surestream Petroleum in 2011. Surestream believes its acreage lies along the western branch of the East African Rift System, including Uganda's productive Albertine Basin[593] and the Rukwa Block in Tanzania. Since the award, Surestream has been conducting an environmental survey to establish what impact drilling could have in this freshwater lake.[594]

In 2012, in response to Malawi's award of Blocks 2 and 3 to Surestream, Tanzania called on Malawi to stop exploring for oil and gas in Lake Malawi (known as Lake Nyasa in Tanzania) until a long-standing border dispute between the two countries can be resolved.[595] Both nations agreed to have the African Forum mediate their dispute,[596] but Malawi President Joyce Banda reportedly said in April 2013 that her country was ready to abandon mediation and take the boundary dispute to the International Court of Justice for arbitration.[597] Malawi claims sovereignty over the entire

Lake based on the 1890 Anglo-German Hligoland Treaty, while Tanzania says it is entitled to 50 percent of it based on the 1982 UN Convention on the Law of the Sea—a battle reminiscent of the dispute between Uganda and the DRC following Uganda's discovery of oil on the shores of Lake Albert.[598] In January 2012, the Tanzanian government awarded the Kyela Block in the northern onshore area of Lake Malawi/Nyasa to the UK's Heritage Oil & Gas, and the company continues to analyze 3D seismic data to select targets for a drilling program.[599]

In July 2012, the government also awarded a 100 percent interest, exclusive prospecting license (EPL 0360/12) for shale gas and coal-bed methane to Australia's NuEnergy covering 346 km^2 in southern Malawi near the Mozambique border.[600] In May 2013, the firm launched its airborne geophysical survey of the depth and structure of sediments in the Mwanza River Basin.[601]

Mali.

Mali is largely unexplored for hydrocarbons and has no established oil or gas reserves.[602] Around one-half of the country overlies sedimentary basin, but only five exploration wells have been drilled—i.e., one well per 450,000 km^2 of basin.[603] The Malian government has identified five promising basins: the Taoudeni Basin (which extends into Mauritania), the Nara Trough (or Nara Graben), the Chad-like Gao Graben, the Niger-like and Libya-like Lullemenden-Tamesna Basins, and the Algeria-Libya Active Oil and Gas Provinces.[604]

Further exploration in northern Mali is unlikely until the country's political situation stabilizes in the aftermath of three events: a March 2012 coup overturning a series of democratically elected governments; the April 2012 takeover of the north by AQIM and its Toureg allies; and the January 2013 counterattacks by France and its African allies attempting to restore democratic, central government control. Reflecting concerns about political risk, Italy's ENI handed back to Mali's government, in January 2013, its licenses to explore the north, citing poor prospects. ENI had acquired five Mali licenses in partnership with Algerian NOC Sonatrach in 2006.[605] Also directly impacted in the north are licenses held by Angola's Petro Plus,[606] Canada's Simba,[607] Qatar's Sphere Petroleum (95 percent interest operator of Blocks 8 and 10),[608] and Venezuela's PDVSA.[609]

This being said, it is possible that exploration work may continue cautiously in southern Mali, which has not been affected directly by the terrorist/secessionist groups. Illustrative of this in a different resource sector is Canadian gold company Iamgold, which said in January 2013 that its operations in southern Mali had been unaffected by recent political events, but that it would "reduce its exploration activity in the region at this time as a precautionary measure."[610] Two IOCs active in the **south** are Canada's Petroma, which has Blocks 17 and 25 north of the capital Bamako,[611] and Bermuda-based AFEX Global, which has carried out 2D seismic work in Block 13 north of Petroma's concessions and confirmed that its Block overlies the promising Nara Graben, a series of Cretaceous rifts.[612]

Mauritania.

Mauritania is a frontier exploration country where production has been disappointing so far, but where promising offshore and onshore prospects have attracted Asian, Australian, European, and U.S. oil firms. The highest profile investor in Mauritania has been Malaysia's Petronas, which bought out the 47.4 percent share in the offshore Chinguetti oil field owned by Australia's Woodside Petroleum in 2007. Petronas is also the operator on the Banda (gas), Tiof (oil), and Tevet fields (oil and gas).[613] Woodside discovered the Chinguetti offshore field in 2001 and started production in 2006 with an estimated reserve of 123 million barrels. Production climbed to 75,000 bpd by late 2006, but "geological difficulties" caused production to drop dramatically to 4,000 bpd in 2011.[614] The Banda offshore field, discovered in 2002, was declared commercial in 2012, and will supply gas to fire an electrical power plant near Nouakchott. The Tiof Field, discovered in 2003, is expected to hold 350 million barrels of oil, while the Tevet Field, discovered in 2005, is believed to contain 50 to 100 million barrels of oil.[615] In 2005, the Faucon Field (gas) was discovered in the south, not far from the maritime border with Senegal.[616]

In 2006, China's CNPC found hydrocarbons in offshore Block 20,[617] for which it has a 65 percent operator interest, with Australia's Baraka Petroleum holding 35 percent.[618] Korea's Dana Petroleum discovered natural gas at the Cormoran exploratory well in 2010 and validated the Pelican gas discovery made 2 km north in 2003, both in Block 7. Dana has a 36 percent share in Block 7, along with France's GDF Suez (27.85 percent), Anglo-Irish Tullow Oil (16.20 percent), and two Australian firms, PC Mauritania and Roc Oil (15 percent and 4.95 percent, respectively).[619] In 2011, Anglo-Irish Tullow Oil appraised the Pelican gas field and discovered yet another underlying field, Petronia.[620] The firm plans a four-well campaign in 2013, starting at the Fregate-Scorpion Prospect which could hold 294 million BOE.[621]

U.S. firm Kosmos acquired a 90 percent interest in three offshore blocks in 2012, with the remaining 10 percent share held by SMH, Mauritania's NOC. Kosmos, which will start drilling in late 2014, believes these blocks are promising because they reside in the proven offshore Mauritania salt basin and include cretaceous geological formations that have been productive in Ghana.[622] Similarly, the UK's Chariot Oil & Gas acquired a 90 percent share in another offshore block in 2012 where petroleum systems are proven and plans to drill in 2014. The firm notes that Mauritania lies in the Central Atlantic Margin, which opened up at same time as the Gulf of Mexico, and also stressed that they contain the same geological source rocks as found in Ghanaian oil fields.[623] In 2012, France's Total acquired deepwater Block C9, which geologically is in the Abrupt Margin in which the firm recently made major discoveries in French Guyana, and plans to drill in 2013 or 2014.[624]

Onshore, exploration continues in the Taoudeni Basin, in central-eastern Mauritania by China's CNPC,[625] France's Total, Germany's RWE,[626] and Spain's Repsol.[627] Total's partners include Algeria NOC Sonatrach and Qatar Petroleum International.[628] Even if commercial discoveries are made, however, the current political uncertainty caused by the presence of AQIM in neighboring Mali may deter onshore development, as adherents of this group have also carried out attacks in Mauritania.

Mauritius.

Mauritius has no known oil or gas deposits. The government has engaged in talks with other nations, including India in 2003 and 2007, about concluding Memoranda of Understanding for cooperation in offshore oil and gas exploration. Nearly 70 percent of the Mauritius population is of Indian origin, and the two nations have a special relationship because of this.[629] Indian oil company ONGC has sought oil and gas exploration rights in Mauritius.[630]

Morocco.

Compared to its prolific oil and gas producing eastern neighbor Algeria, Morocco has had a modest upstream industry.[631] Morocco has a variety of liquid and dry accumulations "ranging from dry gas in the Gharb (also known as Rharb) Basin, condensate in Essaquira, light oil in Essaquira and Prefi, to heavy oil in Tarfaya."[632] The most important oil and gas fields currently in production are the Essaouira Basin on the coast, producing oil and natural gas, and the Gharb Basin in the north. Ireland's Circle Oil is drilling six wells in Sebou and Lalla Mimouna, in the Gharb region[633] south of Tangiers and near the Strait of Gibraltar. An important gas field was also discovered in recent years at Meskala, just north of Essaouira, where condensate has been produced since 1987.[634] There are even unconventional oil shale formations being developed by Ireland's San Leon Energy onshore in southwest Morocco at Tarfaya[635] and at Tmahdit in the middle Atlas Mountains.[636]

Despite the E&P, most sedimentary basins in Morocco onshore and offshore are still largely unexplored. Several factors created enthusiasm among IOCs for Morocco's oil and gas potential, particularly in 2012 and so far in 2013: the government's stability, its competitive taxation regime, a supportive regulatory framework; and a capable NOC partner, ONHYM.[637] Leveraging its deepwater expertise, U.S. giant Chevron acquired rights in January 2013 to 75 percent of three offshore areas in Morocco: Cap Rhir Deep, Cap Cantin Deep, and Cap Walidia Deep.[638] Moroccan NOC ONHYM retained 25 percent shares for both the Pura Vida and Chevron deals. U.S. firm Kosmos has acquired operator shares in four offshore blocks, three in the Agadir Basin, and the fourth in the Aaoun Basin. It plans exploratory drilling as early as the second half of 2013.[639] Kosmos increased its equity shares in two of these blocks and now has 56.25 percent participating interest in the Foum Assaka Block and a 75 percent interest in the Essaouira Block.[640] In January 2013, U.S. firm Plains Exploration and Production (PXP) purchased a 52 percent operator interest in the Mazagan permit from Pura Vida. PXP expects to drill the first of five wells in 2014 in the Toubkal Prospect, which has a mean resource potential of 1.5 billion barrels.[641]

Current IOCs active offshore and/or onshore in Morocco include Australia's Pura Vida Energy, Tangiers Petroleum, and Triangle Energy; Portugal's Galp Energia, and the UK's Caithness Petroleum, Cairn Energy, Chariot Oil, Genel Energy, Gulfsands Petroleum, Longreach Oil & Gas, and Serica Energy.[642] In 2010 and 2011, Spain's Repsol acquired four offshore and onshore exploratory licenses at Tangers-Larache, Gharb Offshore, Hauts Plateaux, Boudenib.[643] New exploration is taking place offshore within the Atlantic Basin and the Mediterranean Sea at Casablanca Offshore, Loukos Offshore, Rabat Deep Offshore, Gharb

Offshore, Tanger-Larache, and onshore at Boudenib, Fes, Foum Draa, Haut Plateaux, Oulad, Safi, Sidi Moktar, Sidi Moussa, Taounate, Tarfaya, Zag, and N'zala.[644]

Mozambique.

Mozambique has two main sedimentary basins: the Rovuma Basin in the northeast adjacent to the Tanzanian border and the Mozambique Basin in the south.[645] Mozambique estimated that it may receive $50 billion in investment by IOCs to develop the country's new gas fields,[646] which are concentrated in Rovuma.

Offshore.

Since signing an oil and natural gas Exploration and Production Concession Contract (EPCC) with Mozambique in 2006 for Area 1 in the Rovuma Basin, Anadarko and its partners have safely drilled more than a dozen deepwater wells within the Offshore Area 1 Block, discovering an estimated 35 to 65+ tcf of recoverable natural gas (approaching 100 tcf of original gas in place [OGIP]). Six of the discovery wells in the Prosperidade and Golfino/Atum compexes (Windjammer, Barquentine, Lagosta, Camarão, Golfhino and Atum) were among the largest discoveries in all of Africa during 2010, 2011, and 2012, according to IHS Energy. Additionally Anadarko and its partners are now advancing a commercial LNG development that is initially expected to consist of four liquefaction trains (5 million tons per annum [mpta] each).[647] Anadarko is now evaluating a third discovery in the block, Tubarao, and will drill an appraisal well there in 2013. Overall, Area 1 is among the world's most significant hydrocarbon finds in the last decade.[648]

Early in 2013, Anadarko and ENI reached an agreement as operators to conduct separate, yet coordinated offshore development activities, while jointly planning and constructing common onshore liquefaction facilities in the form of an LNG park in the Cabo Delgado province of northern Mozambique[649] with a capacity of approximately 50 million tons of LNG per year—enough to elevate Mozambique to one of the world's largest LNG exporters.[650] Reflecting its need to seek massive financing to develop these fields and its share of the LNG park, Anadarko said that the firm was considering selling up to one-third of its interest in the block.[651] NOC Oil India and India's ONGC reportedly made a joint offer for a 20 percent stake in Offshore Area 1.[652] In March 2013, Anadarko was also talking with potential Japanese off-take customers for its LNG production.[653] Anadarko operates Area 1 (36.5 percent interest) with Japan's Mitsui (20 percent), India's BPRL Ventures and Videocon (10 percent each), Thailand's PTTEP (8.5 percent), and Mozambique's NOC ENH (10 percent).[654] The presence of Mitsui, a Japanese general trading company (*sogo shosha*), as a partner makes it more likely that the consortium will ultimately find Japanese off-take customers, and that Anadarko will not sell out its share to a Chinese NOC.

Like Anadarko, Italy's ENI has also enjoyed great success in the Rovuma Basin. ENI made a new discovery in December 2012 in the Mamba Complex of the Offshore Area 4, which is due east of Area 1,[655] bringing its estimate for gas in place at 75 tcf. Its operator interest is 70 percent, with 10 percent each for Portugal's Galp Energia, South Korea's KOGAS, and ENH.[656] In March 2013, ENI reached an agreement to sell Chinese NOC CNPC a 28.6 percent share of its ENI East Africa subsidiary for $4.21 billion,[657] and therefore a 20 percent interest in Area 4.[658] The presence of CNPC and KOGAS make it more likely that Asian off-take customers for this consortium's gas production will be from China and South Korea, respectively.

Due south of offshore Areas 1 and 4 are offshore Areas 2 and 5, still in the Rovuma Basin, and operated by Norway's Statoil (40 percent share) with Japan's Inpex (25 percent), Anglo-Irish Tullow Oil (25 percent), and ENH (10 percent).[659] The partners plan to drill two wells in 2013 2Q.[660] In the author's view, Statoil selected Inpex as a strategic partner, in April 2013, to facilitate long-term gas contracts with Japan. Due south of Areas 2 and 5 are Offshore Areas 3 and 6, operated by Malaysia's Petronas (50 percent share), with French partner Total (40 percent) and ENH (10 percent). "The oil potential of the southern part of the Rovuma Basin might equal the gas potential of the northern part," a Total spokesman has asserted,[661] suggesting that Areas 2, 3, 5, and 6 could be the site of future large discoveries.

Onshore.

Mozambique currently has four onshore fields: Pande, Temane, and Buzi-Divinhe,[662] all discovered in the 1960s, and Inhassoro, discovered by South Africa's Sasol in 2003.[663] Starting in 2004, Sasol exported most of its output through an 865-km pipeline to its South African chemical plants. It currently plans to increase production to supply its expanded synthetic fuels plant in South Africa and a new gas-fired power plant in Mozambique. Production of natural gas from the Pande and Temane fields, operated by Sasol, increased to about 3.1 bcm in 2010.[664] In July 2012, Sasol was analyzing the feasibility of developing the condensate deposit, with a potential of 215 million barrels, which it recently discovered in Inhassoro.[665]

Anadarko is lead operator (35.7 percent) of Rovuma Onshore on the Tanzanian land border adjacent to the Indian Ocean, with France's Maurel et Prom (27.7 percent), ENH (15 percent), the UK's Wentworth (11.6 percent), and Thailand's PTTEP (10 percent).[666]

Further opportunities exist onshore in the provinces of Gaza, Inhambane, Sofala, Zambezia, Nampula, and Cabo Delgado.[667] Several oil companies are also carrying out exploration activities in the Urema Graben in central Mozambique, which is the southern part of the East African Rift System.[668]

Governance Challenges.

Excluding onshore natural gas reserves and the potential for unconventionals such as coal-bed methane, Mozambique may already have 175 tcm of gas between the discoveries to date in Areas 1 and 4. By one estimate, Mozambique, which already grew by 7.9 percent annually from 2001 to 2011, may double or triple its GDP over the next 20-30 years. Yet, just as geological risk for Mozambique is declining, "the political, regulatory, and infrastructure risks are increasing," likely leading to delays in the 2018 start-up for bringing new offshore gas fields to production.[669] On the surface, Mozambique is **politically** quite stable, with the same ruling party, Frelimo, in power and winning by big margins at the polling place since independence from Portugal in 1975. However, a gas and coal boom has raised public expectations in a country where income inequality and poverty level have not improved despite rapid economic growth. There were two bouts of urban unrest in Maputo in 2008 and Tete

194

Province in 2012. Also problematic are **regulatory** gaps in existing oil and gas legislation. One academic felt that the Mozambique government:

> urgently needs to clarify its broader development agenda for the gas sector and make crucial decisions around infrastructure, local gas pricing, capital gains taxes, corporate and social responsibility standards, and social fund payments.[670]

In December 2012, the government was considering levying a 32 percent tax on future sales of local assets,[671] which could discourage farm-outs to and risk-sharing with Asian IOCs. Also discouraging future farm-outs is a proposal to increase in the government's share (though NOC ENH) in oil and gas blocks to a maximum of 40 percent, up from the current of 25 percent.[672] The Mozambique government announced in May 2013 that it expected its new petroleum law to be ratified by year-end 2013 in time for a new licensing round to promote exploration in offshore areas 4, 5, and 6. The government also claimed that it was moving closer to developing a master plan for the development of its new natural gas discoveries, including asset development options, optimal locations, pricing structures, and social improvements. Besides the LNG park, options include fertilizer, petrochemical, gas-to-liquids, and power generation.[673]

On a positive note, in December 2011, the Mozambique government did sign an agreement with Comoros and Tanzania, demarcating their land and maritime borders.[674]

Namibia.

Offshore.

Namibia has attracted investment by major IOCs in 2012, including the UK's BP,[675] Portugal's Galp,[676] and Spain's Repsol,[677] on the bet that its coastal shelf may mirror that of Brazil across the Atlantic Ocean, where the Tupi discovery in 2007 was the biggest find in the Americas in 3 decades.[678] Namibia and Brazil are on the opposite sides of the South Atlantic Margin and share geological history—leading one observer to assert boldly that "Namibia is due to become one of the largest energy producers in Africa [and] might be positioned to rival its neighbor Angola and Algeria in North Africa."[679]

There are four offshore basins in Namibia that are the focus of interest: the Namibe Basin in the north, which starts south of the Angolan border and part of the prolific West African "salt basin." Prior to the Atlantic Ocean opening, the Namibe Basin lay adjacent to the Santos Basin of Brazil, where super-giant oil discoveries have been made. The central Luderitz and Walvis Basins are virtually unexplored, with only four wells drilled as of the end of 2012. To the south, the Orange Basin includes the large delta of the Orange River system and holds the Kudu gas field,[680] which was discovered by Chevron in 1973,[681] and whose production will be exported to South Africa,[682] or sold to NamPower, a Namibian utility.[683] Significant new exploration is being undertaken in the Kudu field in a joint venture led by Anglo-Irish operator Tullow Oil, with Russia's Gazprom, Japan's trading house Itochu, and Namibia's NOC.[684] Brazil's HRT Oil holds 10 offshore blocks in four licenses and drilled its first exploratory well in License 23 offshore in Walvis Bay in March 2013,[685] with a second planned for June 2013. Although HRT did not find oil in commercially viable quantities, the firm indicated that the well "confirms the oil potential of the basin" and "most thought Namibia would only be a [natural] gas potential area."[686] Other companies exploring for oil and gas offshore include Australia's Global Petroleum[687] and Pancontinental,[688] Canada's Eco Oil & Gas[689] and Westbridge Energy,[690] France's Maurel et Prom,[691] Nigeria's Oranto,[692] South Africa's Signet Petroleum,[693] the UK's Serica Energy[694] and Tower Resources,[695] and U.S. firm EnerGulf Resources.[696] Pancontinental announced that it had received an independent Mean Prospective Resource estimate for its EL0037 petroleum license of 8.2 billion barrels of oil.[697] Westbridge, which has an 80 percent share in 1811B and is acquiring a 75 percent interest in adjacent 19-series blocks from Nambian private-sector company Ropat Petroleum Investments, believes that its licenses:

> show similar geologic play concepts and trends to those observed in other prolific petroleum basins in Brazil and the deepwaters of Ghana, Angola, Liberia and Sierra Leone.[698]

Global, which holds an 85 percent interest in PEL-29, has noted evidence of hydrocarbons, including oil slicks on the ocean floor, seabed cores, and 2D seismic, and is believed to have two prospects with greater than one billion barrels each of recoverable oil.[699]

In 2011, the Ministry of Mines and Energy announced that a study had identified the potential for some 44 billion barrels of oil offshore,[700] and claimed that 11 billion barrels in oil reserves had already been found.[701] He predicted first production by 2015, a date that seems likely to slip in the light of three dry wells in 2012, one by Petrobras in the southern Orange Basin[702] and two by Chariot, one in a southern block and the other in a northern block.[703]

A major reason for a lack of exploration offshore has been the significant, expensive challenge of the depths at which prospective fields are thought to be located. In Angola, to the north of Namibia, initial offshore success was in much shallower waters and paved the way for majors to go after promising structures in deeper water.[704] By contrast, in Namibia, there had only been 15 offshore wells ever drilled as of 2011 3Q, meaning that the geology is less well known. Chariot's failed effort to drill in the northern block at the Tapir Prospect to 4,879 meters in 3,123 meters of water reportedly cost the company $65 million.[705]

Onshore.

Three U.S. firms, Duma Energy,[706] Frontier Resources,[707] and HydroCarb, have encountered promising signs of hydrocarbons in the Owambo (Etosha) Basin in north-central Namibia bordering on Angola — one of the largest, unexplored onshore basins in Africa.[708] HydroCarb's COO believes that the "Owambo Basin has all the ingredients to become a major new hydrocarbon province."[709] Frontier Resources has Blocks 1717 and 1817 in the eastern sector of the Owambo Basin, where Occidental Petroleum found evidence of a petroleum system in 1993-94.[710] Frontier believes that the East African Rift System has an extension into northeast Namibia.[711] Smaller independent oil companies such as the UK's Chariot Oil & Gas and Tower Resources, Brazil's HRT, and Canada's Eco Oil & Gas have licenses with shale and coal-bed methane potential, particularly in the Nama/Karoo Basin,[712] which extends into Botswana and South Africa.

Niger.

Niger has three distinct areas of oil and gas potential:

1. Agadem Block: Currently the country's only producing block. It covers the southern half of the Termit-Tenere Rift, bordering Chad to the east. The Niger government estimates the proven oil reserves in Agadem at 324 million barrels, but has also reported that oil reserves could prove to be three times higher than this number.[713]

2. Tenere Block: Comprises the northern part of the Termit-Tenere Rift, which is one of several rift basins that extend across north-central Africa with similar oil-producing basins found in Libya, Chad, and Sudan. Thought to possibly hold 1-3 billion barrels of oil.[714]

3. Kafra Block: A continuation of the rift basin of the Tenere Block; situated in northern Niger up to the Algerian border.[715] Algerian NOC Sonatrach is exploring in this block.[716]

In Agadem, China's CNCP took over in 2003 from an ExxonMobil/Petronas joint venture that was not renewed, taking a 100 percent share in Block Bilma and an 80 percent share in Block Tenere, along with a Canadian oil company, TG World, which took 20 percent. In 2008, Niger and CNCP signed an integrated upstream/downstream package concerning exploration and development of the Agadem Block, and construction and operation of both a long-distance pipeline and a joint venture refinery. The integrated project became operational in 2011 with a 1 metric ton per year (mt/y) oilfield, a 1 mt/y (20,000 bpd) refinery near Zinder, and a 462.5-km oil pipeline to connect the oilfield and refinery.[717] In an interesting example of warming cross-Straits economic ties, CNCP agreed in April 2013 to a joint exploration agreement for the Agadem Block with Taiwan's NOC, CPC.[718]

Instability in neighboring Mali to the west notwithstanding, IOCs are still interested in Niger, particularly in the eastern part of the country. In 2011, Russia's Gazprom reportedly was going to undertake both mining and oil/gas exploration in the Agadez region.[719] In November and December 2012, Australia's International Petroleum[720] and Canada's Brandenburg Energy[721] picked up four leases each.

Chad-Cameroon Pipeline.

In a move to ensure an export outlet, the Niger government signed an agreement with Chad in July 2012 to construct a 600-km pipeline linked to the Chad-Cameroon pipeline. Niger is expected to begin producing oil at four fields in its Agadem Block by early 2014, with production increased to 80,000 bpd, 20,000 of which would be processed at the refinery near Zinder and 60,000 bpd exported through the pipeline.[722]

Nigeria.

The outlook for OPEC member Nigeria, which is both Africa's largest oil producer and holder of its largest oil reserves, hinges on resolving a number of major issues, including civil conflict in producing regions; improving the transportation infrastructure, whose deteriorated state has led to sporadic disruptions to supply; the inability of the Nigerian National Petroleum Company (NNPC) to fund its share of investment in joint ventures with IOCs;[723] and delays in agreeing to a new Petroleum Industry Bill, which is holding back $40 billion in investments in the energy sector.[724] ConocoPhillips' pessimistic view of Nigeria's investment environment was a significant factor behind the $1.79 billion sale of its Nigeria assets in January 2013 to Oando Energy Resources, a Nigerian company.[725] Similarly, Chevron's May 2013 decision to sell its 40 percent undeveloped stake on OMLs 83 and 85 has to be seen as a divestment vote of no confidence, given the firm has put off a decision to develop them despite estimates that two fields, Anyala (in OML 83) and Madu (in OML 85) are thought to have combined reserves of about 250 million barrels and 500 bcf of gas.[726]

Despite new output from deepwater fields, Nigeria's production declined slightly in 2012 from the previous year as increased oil theft[727] and flooding cut crude oil production in the fourth quarter by about 200,000 bpd. Incidents of oil theft and pipeline vandalism continue to curb production in 2013. Crude oil output averaged 2.0 million bpd in the first quarter of 2013, which is 120,000 bpd lower than the same time period last year, despite new production coming online. This is also well below the country's 3.0 million bpd capacity.[728] EIA expects crude oil production to remain relatively flat in 2013 compared to the previous year, although one investment house predicted an increase to 2.3 million bpd.[729] Nigeria's oil industry is primarily located in the Niger Delta, where local groups seeking a share of the oil wealth often attack the oil infrastructure and staff. Current exploration is mainly focused on deep and ultra-deep offshore.[730] The Bonga Field (OML 118) is Nigeria's first deepwater oil field; it began production in 2003 and is responsible for about 600,000 bpd out of the country's 2.0 million bpd in production. Other major oil and gas fields include Akpo, located off the coast of Port Harcourt, and Agbami, located offshore in the central Niger Delta.

In 2011, Anglo-Dutch Shell had the largest daily production among IOCs (974,000 bpd), followed by ExxonMobil (800,000 bpd, of which 700,000 bpd is crude), Chevron (516,000 bpd),[731] and Total (which produced 179,000 bpd before selling some of its interests to Chinese IOCs). By another calculation, Shell (and its joint venture partners, including NOC NNPC) accounted for 26 percent of Nigeria's crude oil production in 2011, while the comparable figures for ExxonMobil were 25 percent, Chevron 22 percent, and Total 12 percent,[732] all including the shares of their respective partners.[733] The remaining unaffiliated firms only accounted for 15 percent of Nigeria's oil production in 2011. One example of several smaller U.S. independents with operations in Nigeria is CAMAC Energy, which has an interest in a PSC for OML 120 with operator ENI of Italy.

China's CNOOC and Sinopec have managed to acquire 209,000 bpd production in Nigeria by buying producing assets rather than participating in bid rounds, making China the fifth-largest IOC producer in the country after Shell, ExxonMobil, Chevron, and Total.[734] CNOOC and Sinopec did this through a series of acquisitions starting in 2006 with CNOOC's $2.268 billion purchase of 45 percent of OML 130, which is operated by Total, and CNOOC's purchase of 38 percent of Stubb Creek (OPL 229), a 15 million barrel field owned by Nigeria's Universal Energy. In 2008, Sinopec took over the operatorship of NPDC's OML 64 and 66, and in 2009 took over Addax, the largest international independent operating in Nigeria.[735] As mentioned earlier, in November 2012, Sinopec also purchased a 20 percent stake from Total in the Usan Field (OML 138/ OPL 222), which started production in 2012. CNCP used a different approach to enter Nigeria. In 2006, CNPC reached an oil and gas cooperation agreement with the government of Nigeria and won the tender of four blocks, namely OPL298, OPL471, OPL721, and OPL732. Block OPL298 and Block OPL471, of which one is onshore and the other offshore, are located in the Niger Delta. Block OPL721 and Block OPL732, the latter a risk exploration block, are located in the Chad Basin in Borno Province, northern Nigeria.[736] As discussed previously, China also indirectly acquired Nigerian oil and gas assets through its acquisitions of two Canadian energy companies, Addax and Nexen. Among other BRIC countries, India's ONGC Videsh had shares in deepwater Blocks OPL 279 and OPL 285 (45.5 and 64.33 percent operator shares) and a 13.5 percent non-operator share in the Nigeria-Sao Tome & Principle JDZ Block 2.[737] Oil India had a 17.5 percent non-operator share in onshore OPL 205.[738]

Nigeria is notable in having a significant number of indigenous oil companies, including SAPETRO and Oando, which also have operations in other African countries.[739] Part of the reason for this phenomenon is the Nigerian government's policy of encouraging foreign companies to establish partnerships with local Nigerian oil companies under the indigenous and marginal field programs. Canada's Mart Resources is an example of a small IOC independent which has created an investment niche for itself in Nigeria under this program.[740]

The following oil projects are planned to come on-line in the next few years, although several are likely to be delayed pending passage of the Petroleum Industry Bill and resolution of other regulatory issues:

Project	Capacity ('000 bpd)	Est. Start-up	Sector	Operator
Agbami	100	2011-14	deepwater	Chevron
Ebok (phase 2)	35	2012	offshore	Afren
Gbaran Ubie	70	2012+	onshore	Shell
Ehra North (phase 2)	50	2013+	deepwater	ExxonMobil
Oberan	TBD	2013+	deepwater	ENI
Ofon (phase 2)	90	2014	offshore	Total
Aje	TBD	2014	deepwater	Yinka Folawyo
Bonga North, Northwest	50-150	2014+	deepwater	Shell
Bonga Southwest and Aparo	140	2014+	deepwater	Shell
Egina	150-200	2014+	deepwater	Total
Bosi	135	2015	deepwater	ExxonMobil
Nsiko	100	2015+	deepwater	Chevron
Uge	110	2016	deepwater	ExxonMobil
Nkarika	TBD	2019	offshore	Total
Etan/Zabazaba	110	TBD	deepwater	ENI
Okan	25	TBD	offshore	Chevron

Source: "Nigeria," *EIA Country profile*, October 16, 2012.

Table A-I. Upcoming Oil Projects in Nigeria.

New output from deepwater fields could increase production in 2014, but this depends on unplanned disruptions, which could continue to offset new production gains, and whether projects come on stream as scheduled.[741] A group of IOCs have said that Nigeria's plan to achieve 4 million bpd production and 40 billion barrels in reserves by 2020 is unrealistic under the new fiscal terms being proposed by the federal government in the draft Petroleum Industry Bill being considered by Parliament.[742] One prediction by respected oil industry consultants Wood Mackenzie indicated that Nigeria's oil production could fall to 1.84 million bpd by 2020 because of aging onshore fields, and insufficient investment in new offshore projects to compensate due to a negative investment climate.[743] Despite this pessimism, ExxonMobil has decided to press on with its Erha North offshore project.

In addition to oil, Nigeria holds the largest natural gas reserves in Africa at 180 tcf (as of the end of 2011), but has limited infrastructure in place so far to develop the sector and only produced one tcf of dry natural gas in 2011.[744] A major Nigerian government priority is to advance the Brass LNG project,[745] the second LNG project after the NLNG project, which was started in 1999 and has Nigeria's NOC as its operator. The government has been working for several years to end natural gas flaring, but the deadline to implement its 2009 Gas Master Plan has been delayed repeatedly, in significant part because security risks in the Niger Delta have made it difficult for IOCs to construct infrastructure that would support gas monetization.

A significant portion of Nigeria's marketed natural gas is processed into LNG. In 2010, Nigeria exported 875 bcf of NLG, making it the fifth largest LNG exporter in the world. Nigeria also began to export some of its gas in 2010 via the West African Gas Pipeline to neighbors to the west: Benin, Togo, and Ghana. The pipeline is owned by Chevron (36.7 percent), Nigeria's NOC (25 percent), and Shell (18 percent), with the remainder held by state-owned shareholders from Ghana, Togo, and Benin. Chevron also operates the Escravos Gas-to-Liquids project, which is scheduled to be operational by 2013.[746]

Inland Offshore.

The Nigeria National Petroleum Corporation (NNPC) announced in September 2012 that it had discovered oil in the Nigerian territorial waters of Lake Chad, which it shares with neighbors Cameroon, Chad, and Niger. USGS estimates that the region of Lake Chad holds 2.32 billion barrels of oil and 369 bcm of natural gas.[747] NOC Oil India has a 25 percent share in OML 142.[748]

Rwanda.

Rwanda has no oil production,[749] but does have methane gas reserves along its borders with the DRC and Tanzania that are being developed. Three oil and gas developments in Rwanda of note are 1) ongoing oil and gas exploration by Canada's Vanoil, 2) an under-construction methane-powered electricity project at Lake Kivu undertaken by U.S. investment firm Contour Global, and 3) an African Development Bank-funded feasibility study that could lead to the construction of a regional pipeline.

1. Exploration in Lake Kivu. Vanoil has a technical evaluation agreement with Rwanda's government covering approximately 1,631 km² in the northwestern part of Rwanda, best known as the East Kivu Graben, located beneath Lake Kivu. Under the agreement, Vanoil must place its first two wells anywhere

within Blocks 3A and 3B by July 31, 2013.[750] The Kivu Graben area is part of the East African Rift System and is approximately 90 km wide and 200 km long. The graben straddles both Rwanda and the DRC and is the southern extension of the Albertine Graben in Uganda, where one billion barrels have been discovered by Anglo-Irish Tullow Oil and the UK's Heritage Oil. Vanoil believes that Lake Kivu's common origin (shared tectonic and sedimentary in-fill) with Lake Albert, Lake Edward, and Lake George make this a promising region for exploration. Vanoil, through contractors, has completed both an aerial survey of the earth's gravity and magnetic fields over Lake Kivu and southwest Rwanda, and a 2D marine seismic survey.[751]

2. Methane-Powered Electricity. Kivuwatt Ltd., a subsidiary of sustainable development-focused Contour Global of the United States, is building facilities in Rwanda to extract natural gas from Lake Kivu and pipe it to a new gas-fired power station. State-owned Kubuye Power already extracts methane gas from the lake, which is used to feed a 4.2 megawatt power station.[752] In 1936, naturally occurring methane was first discovered in Lake Kivu, which in 1976 was estimated to contain a 55-60 km^3 field with an increasing quantity of trapped methane gas. In the project's first phase, processed methane gas will produce approximately 25 megawatts (MW) of electricity; phase 2 is expected to add another 75 MW of capacity, all for the local Rwandan power grid.[753]

3. Regional Pipeline. The East African Community (EAC) Secretariat has announced that it has funding for a pipeline south from Rwanda to Burundi, but needs ADB funding and support for a second, new pipeline linking Rwanda to an existing pipeline be-

tween Kampala, Uganda, and Eldoret, Kenya. A feasibility study carried out in 2007 indicated that the Kigali-Kampala pipeline was commercially viable and would cost $193.6 million to build.[754]

Sao Tome and Principe.

To date, there have not been any major commercial oil or gas finds in Sao Tome and Principe, but there is guarded optimism about future discoveries. Three oil and gas stories for Sao Tome and Principe concern 1) E&P in the Joint Development Zone (JDZ) with Nigeria; 2) E&P in Sao Tome's EEZ; and 3) plans to build a tank farm and oil transshipment port for international use firms operating in the Gulf of Guinea sub-region.

JDZ.

Sao Tome and Nigeria signed a treaty in 2001 to create the JDZ, divided it into nine blocks in 2003, and solicited bids in a first licensing round in 2004. Sao Tome has a 40 percent share in the JDZ, and Nigeria has a 60 percent share. The Joint Development Authority, created under the treaty, awarded U.S. firms Chevron and ERHC exploration rights, respectively, for Block 1 in 2004 and Blocks 2 to 6 in 2005.[755] (ERHC currently also has a 20 percent share in Block 9.) Block 1, located in 1,600 to 1,800 meters in water depth, is the most promising and is operated by France's Total, which purchased Chevron's 49.5 percent stake in 2010 in order to create cost and operational synergies between Block 1 and its stake in the adjacent Nigeria Block 248.[756] Other shares in Block 1 are Chinese owned Addax, South Africa's Sasol, Nigeria's Dangote Energy, and UK-firm Afren/DEER. Obo-1, drilled in 2006 in Block 1, proved a working hydrocarbon system in the JDZ,[757] but Total's Obo-2, drilled in May 2012, was plugged and abandoned.[758] Total continues to evaluate commercial viability of the Obo Field.[759] "At all of the wells so far drilled in the Joint Zone, hydrocarbons have been found—either gas or oil—but the question is if these deposits do or do not have a commercial value," the JDZ's Executive Director said in August 2012. One interesting element of the JDZ has been the Asian presence of China's Sinopec, which has operated Block 2: Addax, a firm bought out by Sinopec in 2009, which operates Blocks 3 and 4; and an Indian oil firm, ONGC Videsh, which has written off its 13.5 percent share in the up-to-now disappointing Block 2.[760] Hong Kong-based private company Sinoangol (not to be confused with China Sonangol) has applied for exploration rights in Block 2.[761]

EEZ.

The EEZ covers an area of 160,000 km² south and east of the JDZ. The STP government launched bidding for 9 of 17 blocks in its EEZ in 2010. Because of limited success in the JDZ, there were no major international oil companies who bid on the blocks, which were awarded, *inter alia*, to smaller African oil producers: Nigeria's Oranto Petroleum, Angola's Grupo Gema, South Africa's Afex,[762] and OG Engineering of STP. The STP government later awarded 100 percent interests to U.S. firm ERHC Energy (in Blocks 4 [zone A] and 11 [Zone B]) and UK firm Equator Exploration (in Blocks 5 [Zone A] and 12 [Zone B]).[763] (Nigeria's Oando Ltd. purchased a majority stake in Equator in 2011.[764]) ERHC believes that the close proximity of STP's offshore waters to the proven hydrocarbon systems in the adjacent territorial waters of Gabon, Cameroon, Equatorial Guinea, and Nigeria suggest the potential for hydrocarbons. The firm believes that the EEZ has "all the necessary components of a successful petroleum system."[765]

Tank Farm /Oil Transshipment Port.

Sao Tome plans to build a refined petroleum tank farm, financed by the Nigerian government. The farm would:

> serve as a refined petroleum transfer center for Nigeria and other Gulf of Guinea ports which suffer from port backlogs. This would allow larger ocean-going vessels to offload at the tank far and for subsequent transfers to final West African destination by smaller vessels or barges.[766]

The STP government signed a contract with Swiss firm Gunvor to construct a $200 million dollar port on the north side of the island of Sao Tome.[767]

Senegal.

The main oil and gas E&P areas in Senegal are in offshore blocks in Senegal's territorial waters or EEZ and in Senegal's JDZ with Guinea-Bissau.

Offshore and Onshore Senegal.

Offshore exploration in Senegal started in the late 1970s, resulting in one field of heavy crude oil and one gas field (Daim Niadio 14). Senegal has 19 blocks, running from Block A (Saint Louis Offshore Deep in the north) to Block S (Casamance Onshore in the south), of which 10 are offshore and 9 onshore. The only hydrocarbon production in Senegal at present is from the Gadiaga Field, about 60 northeast of Dakar, which is shared 70 percent by Houston-based Fortesa and 30 percent by NOC Petrosen. Fortesa also has a license to explore the Tamna (Zone G) onshore block.[768] Other foreign investors in Senegal's oil & gas industry include Nigeria's Oranto, which has PSCs for three blocks.[769] Senegal's total annual crude oil production was 99,000 bbl in 2008, 249,000 bbl in 2009, and 398,000 bbl in 2010.[770]

Anglo-Irish Tullow operates the 2,807 km² Saint Louis offshore block (Zone B) in the **north** with a 60 percent interest. Tullow is optimistic about further exploratory drilling in 2013 because of the results of a seismic survey carried out on the boundary between Senegal and Mauritania in 2008, and an exploratory well drilled in 2011 in the same Cretaceous turbidite play that led Tullow to discover the Jubilee Field in Ghana. This license covers the northern-most inshore section of the Saint Louis offshore area and adjoins Mauritanian Block 1, where Tullow also has nine contiguous licenses.[771] The UK's Cairn Energy (operator) and First Australian Resources (FAR) have a 35-65 percent interest in three **central**, offshore blocks—Rufisque (Zone K), Sangomar Deep (Zone Q), and Sangomar (Zone O)—covering 7,490 km² over the Senegalese portion of the productive Mauritania-Senegal-Guinea Bissau Basin.[772] FAR's OOIP estimate ranges from 200 million bbl to greater than 1 billion bbl potential, in part because of parallels its block has to Mexico's Cantarell super-giant field. In pre-rift time, Senegal was considered to be adjacent to Mexico before the African and American continents pulled apart.[773] The UK's Elenilto operates a 7,920 km² block in the Southern

Senegal Casamance offshore basin in the **south**, with Petrosen as a minority shareholder.[774] Elenilto calculates that there are 500-800 million barrels in hydrocarbon content in this block, which is similar to other fields elsewhere in West Africa (Mauritania, Sierra Leone, Liberia, Cote d'Ivoire, and Ghana).[775]

JDZ with Guinea-Bissau.

The UK's Ophir Energy has a 79.2 percent operated interest in the Profond Block of Senegal's JDZ with Guinea-Bissau.[776] Although Ophir's Kora-1 exploratory well was drilled and abandoned in 2011,[777] Profond is estimated to contain 400 million barrels of oil.[778]

Seychelles.

Although international oil companies have explored for oil and gas in the waters around the Seychelles since the 1970s, no commercially viable hydrocarbons had been found as of year-end 2012. Nevertheless, the government's Seychelles Petroleum Company (Sepec) has aggressively sought foreign investment,[779] including contracting for 20,000 km of seismic data in 2011, which it expected would "significantly expand . . . [the] interest in the hydrocarbon potential of the Seychelles."[780] Global interest in the waters off the coast of the Seychelles has also been sparked following recent success in the region, including Anadarko's large gas discoveries off the Mozambique coast and BG's promising gas discoveries in Tanzania.[781] In November 2012, the Seychelles government delayed a bid round on 30 blocks to give itself time to seek advice from the International Monetary Fund on negotiations with potential investors,[782] including possibly increasing petroleum revenue royalties to 10 percent, from 5 percent.[783] The government plans to revise its draft petroleum law by May 2013, after which the government will invite companies to negotiate on a one-on-one basis for exploration licenses.[784]

There are currently two main foreign oil companies exploring in the Seychelles: Australia's WHL Energy has a 100 percent interest in 20,476 km^2 off the southern coast,[785] and the UK's Afren operates a 75 percent interest in approximately 14,319 km^2 in Areas A and B, which are located in the mainly shallow water in the northern half of the Seychelles plateau. (The UK's Avana Petroleum has a non-operator 25 percent interest.) WHL undertook extensive 2D seismic work and reprocessed existing seismic dates, rebuilding hydrocarbon basin models with reference to other hydrocarbon rich basins in East Africa in particular.[786] The firm hired an independent consultant, which announced in November 2012 that WHL had a total unrisked mean prospective resource of 3.45 billion barrels for the 21 most highly ranked leads in its portfolio.[787] WHL announced in April 2013 that it had signed a letter of intent with a top 100 Fortune Global 500 oil and gas company to farm into its frontier exploration effort.[788]

Afren commenced a major 3D seismic program in Areas A and B in late 2012,[789] and earlier had relinquished its license in Area C in order to focus on these first two areas.[790] The firm believes that the tectonic evolution of the underwater Seychelles Plateau is related to three phases of rifting that isolated the now drowned micro-continent from the center of Gondwanaland.[791] The Western Seychelles Margin, according to Afren, can be reconstructed to a position adjacent to Somalia and as a northern extension of Madagascar prior to drifting from Africa. A second phase of rifting occurred when India and Seychelles separated from Madagascar.[792] Plate reconstructions also show the geological closeness of the proven petroleum systems in both India and Madagascar.[793]

Sierra Leone.

The discovery of Ghana's Jubilee field in 2007 triggered the interest of international oil companies in exploring offshore West Africa in an area known as the West Africa Transform Margin, which stretches 1,500 km between two tectonic plates along the coasts of eastern Ghana, Cote d'Ivoire, Liberia, and Sierra Leone. A hydrocarbon assessment published by the U.S. Geological Survey in 2011 of an area of about 202,700 km^2 in offshore Guinea, Liberia, and Sierra Leone estimate mean volumes of undiscovered hydrocarbon resources in this area to be 3.2 billion barrels of oil, 23.63 tcf of gas, and 721 million barrels of natural gas liquids. Recent discoveries, including oil discoveries drilled by U.S. firm Anadarko Petroleum (Venus in Block SL-06 in 2009, Mercury-1 in Block 07B-10 in 2010, and Jupiter-1 in 2012),[794] indicate the likelihood of major oilfields being discovered offshore in Sierra Leone because they confirmed that the main reservoir and source elements in the Jubilee Field in Ghana are present in the Sierra Leone/Liberia deepwater basin.[795] Exploration remains at an early stage, but prospects look good for Sierra Leone to become a new frontier for oil and gas.[796] Assessments made in Sierra Leone suggest high potential and good possibility of large oil and gas field size (500-700 million barrels of oil and 4 tcf of gas).[797]

Sierra Leone's government awarded eight offshore oil blocks in July 2012, teaming various bidders together as forced partners for four of the blocks. U.S.-based Kosmos Energy and Australia's African Petroleum were awarded Block 4A; South Africa's Signet Petroleum and UK-firms Menexco Petroleum and Elinilto received Block 7a; and the U.S. firms Chevron (operator, with 55 percent interest) and Noble Energy (30 percent), as well as UK firm Odye (15 percent), were awarded Blocks 8a and 8b, which will become one concession.[798]

After earlier awards and farm-ins, Block SL-4b-10 is now operated by Canada's Talisman (30 percent), with other shares owned by Russia's Lukoil (25 percent), Malaysia's Petronas (25 percent), and the UK's Protinal (20 percent).[799] Block 07B-11 is owned by Anadarko, Anglo-Irish Tullow, and Spain's Repsol.[800] Lukoil had announced a $100 million drilling program for its 49 percent stake in another block, SL-5-11, which is also co-owned by Nigeria's Oranto (30 percent) and the U.S. firm PanAtlantic (21 percent).[801] Lukoil and Vanco are partners in a number of blocks in West Africa's frontier acreage, including in Ghana.[802] Repsol acquired three offshore blocks in 2011, which the firm believes "could be a new oil area."[803]

Somalia.

Estimates of onshore and offshore reserves in Somalia vary greatly, but run as high as 110 billion barrels of oil.[804] The country is also known to have significant deposits of natural gas.[805] Oil seeps were first identified by Italian and British geologists during the colonial era. Exploration activities had focused on northern Somalia, where several American firms including

Amoco, Chevron, ConocoPhillips held concessions, but were forced to declare force majeure following the collapse of the central government in 1991.[806] Independent international E&P companies are presently investing in Puntland and Somaliland despite the legal and political uncertainties. Somalia's federal government, which gained more international credibility following the successes of the African Union Mission in Somalia (AMISOM) in beating back the terrorist group Al-Shabaab, which had taken control of large swaths of south-central Somalia. The federal government has stated that it will recognize permits issued prior to 1991, but has been undecided on whether post-1991 concessions, including those in Puntland and Somaliland, would be recognized.[807] In April 2013, Somalia Petroleum Corporation announced that the government intended to sign 30 PSCs by year-end, starting with firms that held acreage before the civil war. This will be complicated, and resolving the situation may take years, not least because several blocks have already been relicensed by Puntland and Somaliland.[808] Challenging the central government, in November 2012, Somaliland's Minister for Mining, Energy, and Water Resources announced an "open door policy" for investors in his region.[809]

Another uncertainty for oil and gas exploration in Somalia has been a dispute with Kenya over the two countries' maritime boundary. The government of Somalia claims that the awards made by Kenya to ENI and Total are illegal because they lie in waters claimed by Somalia. Kenya's position is that the maritime boundary should run due east from the land border, while Somalia says the boundary should extend perpendicular to the coastline.[810]

Puntland.

These optimistic estimates of Somalia's hydrocarbon potential are derived in part from geological analysis suggesting that Puntland's Dharoor Valley in the north and the Nugaal Valley in the south are extensions, respectively, of the Masila and Marib-Shabwa Basins of Yemen, which currently produce 400,000 bpd. The northern part of East Africa was part of the much larger Gondwana mega-continent until Puntland was finally separated from the Arabian Peninsula during the opening of the Gulf of Somalia, at which point the Dharoor and Nugaal Basins were separated from their analogues in Yemen.[811]

Australia's Range Resources estimated that the northern Puntland province had the potential to produce 5-10 billion barrels of oil.[812] These two blocks cover an area of over 9.7 million acres. With only three wells drilled in the two block areas, this is one of the least explored areas in Africa.[813] Signs of oil were confirmed in Dharoor in January 2012, when Canada's Horn Petroleum drilled the first well in Somalia in 20 years in Dharoor,[814] adding a second well in June 2012. News reports about the two wells have been contradictory, with one source indicating that the two wells are thought to contain 300 million barrels of oil each,[815] while another indicated that commercial quantities of oil had yet to be found.[816] Further testing is needed to determine best prospects, but Horn estimates that the Dharoor Field contains 6 billion barrels of oil,[817] and has made commitments to drill at least two more exploratory wells by 2015.[818] Horn Petroleum's share in its two blocks in Dharoor and Nugaal is 60 percent, with 20 percent each for Range Resources and Red Emperor.[819] Red Emperor believes that Dharoor and Nugaal have 19.9 billion barrels of oil in place.[820] Range first entered Puntland in 2005 and has an unofficial role as an advisor to the Puntland administration on attracting international oil and gas investment.[821] The CEO of Canada's Africa Oil, which has a 45 percent equity interest in Horn Petroleum,[822] stated in August 2012 that:

> while we were disappointed that we were not able to flow oil from the first two exploration wells in our Puntland drilling campaign, we remain highly encouraged that all of the critical elements exist for oil accumulations, namely a working petroleum system, good quality reservoirs and thick seal rocks.[823]

Somaliland.

The autonomous region of Somaliland contains geological basins similar to both Yemen and Uganda, where millions of barrels of oil reserves have been discovered. As with the rest of Somalia, these estimates are highly speculative, however, and only a cumulative total of 21 wells had been drilled in Somaliland as of the end of 2012. There may also be vast reserves of natural gas offshore,[824] as appears increasingly to be the case for much of the coast of East Africa, running from Mozambique, Madagascar, Tanzania, and Kenya. The UK's Genel Energy was awarded an exploration license for onshore Blocks SL-10-B and SL-13 in Somaliland, with a 75 percent interest in both (along with 25 percent share partner East Africa Resource Group), as well as farmed-in for a 50 percent share in the Odewayne PSA, which covers Blocks SL-6, 7, and 10A (along with 30 percent for Australia's Jacka Resources and 20 percent for Petrosoma).[825] Genel plans a 2D seismic campaign for its blocks in 2013, with the first exploration well in the second half of 2014.[826] Genel believes the block could contain one billion barrels of prospective resources, with 15 percent probability of success.[827] The UK's Ophir has a 75 percent operated interest in Berbera Blocks (SL 9 and SL 12),[828] while Norway's DNO International was awarded onshore Block SL 18.[829]

South Africa.

South Africa's proven oil reserves total a trifling 15 million barrels,[830] but this figure could increase dramatically in the coming years. Besides the modest and declining reserves offshore Mossel Bay in eastern South Africa, the country's only other current reserves are offshore in southern South Africa in the Bredasdorp Basin and in western South Africa, particularly in the Orange Basin adjacent to Namibia.

Offshore exploration in South Africa is intensifying, spurred by: 1) massive gas discoveries to the east in Mozambique in 2010, 2) modest discoveries in the Bredasdorp Basin to the south as the result of drilling since the 1980s, and 3) encouraging results in the Orange Basin in the east, south of Namibia.[831] Onshore, much of the exploration is in unconventional oil and gas in the Karoo Basin in the west near Botswana, and in coal-bed methane in the northeast.

Extending the East Africa Gas Boom.

U.S supermajor ExxonMobil will be the operator for exploration in 2.8 million acres of South Africa's east coast, hoping to expand the energy boom along Africa's Indian Ocean flank that was set off by massive gas discoveries off Tanzania and immediate neighbor

Mozambique. In December 2012, Exxon acquired a 75 percent operator interest in exploration rights in the Tugela south blocks from the UK's Impact Oil and Gas and acquired the right to buy a 75 percent interest in three other offshore blocks for which Impact currently holds technical cooperation permits.[832] The U.S. giant also signed a technical cooperation agreement with the South African government to study oil and gas potential in the deepwater Durban Basin, which covers about 12 million acres of seabed.[833] Sasol holds exploration rights in another large block east of the Tugela permit and exploration right areas, up to the border with Mozambique.[834] Impact has sought a license for technical studies in the Transkei Permit area offshore East London, which is further west of the Tugela blocks. Even further west, in the area offshore of South Africa's southernmost Cape Agulhas, PetroSA operates the small Sabel oil field, and several IOCs have technical cooperation permits and exploration rights, including France's Total, Anadarko, and smaller oil and gas companies, or "minnows," such as Impact, Singapore's Silver Wave Energy, as well as South Africa's PetroSA.

Offshore Oil, Gas-to-Liquid.

South Africa's PetroSA operates the Oribi and Oryx fields, located in Block 9 off South Africa's southern coast in the Bredasdorp Basin. The fields are 120 km southwest of Mossel Bay, where a gas-to-liquid refinery is located. These two fields currently only produce 1,800 bpd, with a declining reserve. PetroSA's Project Ikhwezi involves tapping into gas reserves in the F-O field, which is 40 km from its current offshore production platform. First gas is scheduled to flow in May

2013, with production lasting 6 years and possibly to 2025 with development of other nearby gas fields.[835]

Western Cape Oil & Gas E&P.

The Orange Basin is believed to hold substantial oil and gas reserves, but exploration is in a preliminary stage and large-scale commercial production is years away, with first gas in 2016.[836] Large international oil firms Anglo-Dutch Shell and the U.S. firm Anadarko, joined in other blocks by the UK's Cairn India and Afren, Australia's Sunbird and BHP, and South Africa's PetroSA, Sasol, Thombo, and Sungu Sungu,[837] are betting that the "geological strata running on land and under the sea southward from Uganda and the Great Lakes extend to South African territory."[838] In August 2012, Cairn India, owned by the UK's Vedenta Group, acquired a 60 percent operator share of offshore Block 1 in the Orange Basin along the maritime border with Namibia, with PetroSA holding the remaining 40 percent.[839] Sunbird Energy acquired a 76 percent share in the 870 bcf Ibhubesti Gas Project in Block 2A offshore, with PetroSA holding the remaining 24 percent. "Following the recent launch of PetroSA's Ikhwesi Project, the Ibhubesi Gas Field is now the largest undeveloped gas discovery in South Africa."[840] Thombo is the operator for development of Block 2B, which is located in the Orange River Basin offshore shallow water area lying between the Ibhubesi gas field and the Namaqualand coast. Other partners include the UK's Afren (25 percent share). Thombo started 3D seismic work 2013 1Q[841] and exploration drilling is expected in 2014.[842]

Karoo Unconventional Oil and Gas.

U.S. supermajor Chevron announced in December 2012 that it would team up with Ireland's Falcon Oil and Gas to seek unconventional exploration opportunities in South Africa's semi-desert Karoo Basin in central and southern South Africa.[843] The EIA estimated in 2011 that there were 485 tcf of technically recoverable gas in the Karoo,[844] which contains the world's fifth largest shale gas reserves.[845] Falcon currently has a Technical Cooperation Permit giving it exclusive rights to obtain an exploration permit in the Karoo Basin. Chevron's move came only 3 months after South Africa lifted a temporary ban on shale gas exploration, or "fracking." South Africa had awarded Shell a technical cooperation permit in 2009 to determine the Karoo's natural gas potential, and the firm submitted exploration license requests for the Western, Eastern, and Northern Cape Provinces in 2010,[846] but strong opposition by environmentalists led the South African government to initiate the temporary ban.

Coal-Bed Methane.

Australia's Kinetiko has a 49 percent, operator share with South Africa's Badimo Gas in a coal-bed methane project in eastern Transvaal, where coal mining has occurred for over 100 years. It expects to drill 7-8 exploration wells in its northern (56ER) and southern (38ER) Amersfoort project license areas by June 2013, allowing it to better estimate first gas reserves.[847] Kinetiko has a 100 percent share in five other concessions, while Badimo has one concession. The U.S. consultancy Gustavson Associates calculates P50 Gas in Place (GIP) resource for Badimo's Amersfoort tenements of 1.7 tcf and P50 prospective resources of 1.12 tcf.[848] Sunbird has four coal-bed methane projects in eastern South Africa (gas estimates in parentheses): Somkele, 200 km northeast of Durban (best estimate gas-in-place: 190 bcf), Ermelo, 200 km southeast of Johannesburg (800 bcf), Springbok Flats, 150 km northeast of Pretoria (620 bcf), and Mopane, 420 km northeast of Johannesburg (1.9 tcf).

The South African government also hopes that domestic offshore natural gas and onshore shale gas, coal-bed methane, as well as gas imports from bordering neighbors Namibia and Mozambique, will provide an alternative to carbon-intensive coal, particularly for electricity and synthetic fuels.[849]

Sudan/South Sudan.

After decades of war, Sudan and South Sudan signed a peace agreement in 2005, leading after a referendum to the independence of South Sudan in July 2011. Although now independent, the two countries remain interdependent in terms of the oil industry because about 75 percent of oil production originates in the south, while the entire pipeline, refining, and export infrastructure is in the north. Oil fields in Blocks 1, 2, and 4, collectively known as the Greater Nile Oil Project, are split between the two countries, since it covers an area that straddles the North, South, and disputed Abyei region.[850] A dispute over pipeline and export transit fees lingered post-independence, leading South Sudan to halt exports for several months to pressure the North to reach an accord on fees. A new peace deal was reached in September 2012, including a deal on fees, but implementation has been held up, *inter alia*, by renewed differences over border security. South Sudan had discussed exporting as much as 200,000 bpd via the pipeline through Sudan,[851] but proposed an interim measure in March 2013 of exporting its crude oil by road via Ethiopia to Djibouti.[852]

Sudan and South Sudan averaged about 425,000 bpd production in 2011, a significant decline from 470,000 in January 2011, mainly due to labor shortages. The progressive shutoff in oil exports via Sudan starting January 23, 2012, until at least early 2013 meant there was little production in 2012. In order of importance in 2011, Sudan and South Sudan oil exports went to China (66 percent), followed by Malaysia (9 percent), Japan (8 percent), United Arab Emirates (5 percent), India (4 percent), Singapore (4 percent), and others (4 percent). Sudan and South Sudan had reserves as of January 2012 variously estimated to be between 4.2 billion barrels (Wood Mackenzie) and 6.7 billion barrels (BP 2011 Statistical Review). Before the split, Sudan held 0.5 percent of global reserves and was considered a minor producer with 490,000 bpd in 2009.[853] Most natural gas is flared or re-injected, despite known reserves of 3 tcf.[854]

To strengthen its own independence, South Sudan—the world's most oil-dependent state with 98 percent of revenues coming from crude[855]—also initiated talks with several neighboring countries about alternative options for pipelines to ports in East Africa,[856] and sought financing from China. As part of its immediate negotiating tactics with the north and to ensure future independence from Sudan's pipeline/port infrastructure, South Sudan signed nonbinding memoranda with Kenya in January 2012 to build a port to the Kenyan city of Lamu[857] and sought in February 2012 to strike a deal with Ethiopia to build a pipeline via Ethiopia to Djibouti's port.[858] The Kenya option would cost $3 billion and take at least 3 years to build. Another option would be a pipeline to Uganda, potentially linking up with discoveries there, and further transporting the oil to a port in Kenya.[859] Whatever new pipeline option(s) is/are chosen, the author believes that China, which up to now has been reticent to express its willingness to finance the pipeline out of fear of antagonizing the north, will eventually finance the new pipeline(s), both to benefit from the construction contracts, and to firm up its relations with (and access to oil from) South Sudan. In March 2013, Sudan and South Sudan signed an agreement, mediated by the African Union, which would allow the resumption of at least 350,000 barrels of oil exports from South Sudan.[860] Barring political issues or technical difficulties that may delay the pace of the ramp up and restart at additional fields, EIA expects combined oil output in both countries to average 210,000 bpd in 2013 and 450,000 bpd in 2014.[861]

Sudan signed a new deal with small Finnish company Fenno Caledonian in 2010 for an 85 percent share in Block 10 in the northeast, with Sudapet taking the remaining 15 percent. Fenno Caledonian, which also has a share in Block 14, believes that Block 10 could have oil or gas because Block 8 had gas showings and the block is part of the same sedimentary base as all other blocks in the southern part of Sudan.[862] Khartoum launched bidding in early 2012 for six blocks clearly located in Sudan with the hope that new exploration could lead to production to replace the oil lost to South Sudan and to maturing fields in the north.[863] In November 2012, Sudan's Petroleum Minister visited Brazil to court oil cooperation, including participation by NOC Petrobras.[864] Khartoum also inaugurated the new Hadida Field in South Kordofan State in

December 2012, with initial production of 10,000 bpd expected to grow to 60,000 bpd. Sudan at year-end 2012 was producing at 120,000 bpd.[865] However, South Kordofan, which borders with South Sudan, has been the site of continued instability because of conflicts between rival tribal groups centered on the disputed town of Abyei that has led to the displacement of tens of thousands.[866]

Following sanctions placed on Sudan, Western oil companies left Sudan, allowing foreign companies from Asia, such as the China's CNPC, India's ONGC, and Malaysia's Petronas, to dominate Sudan and South Sudan's oil sectors through various shares in three consortia whose minority shareholder include Sudan NOC Sudapet and/or South Sudan NOC Nilepet.[867] China's Sinopec (6 percent) and Egypt/Kuwait's Tri-Ocean Energy (5 percent) are also minority share-holders in one of the three consortia, Dar Petroleum (Petrodar), which is based in South Sudan.[868] ONGC Videsh's has two projects in Sudan and two in South Sudan.[869] France's Total owns an interest in Block B in South Sudan and had been preparing to resume exploration activities[870] when the South Sudanese government, exercising its right under new hydrocarbons legislation, sought to split the massive concession in November 2012 into three parts,[871] allowing Total to keep one part and offering the other two to ExxonMobil and Kuwait's NOC Kufpec.[872] Total preempted the move in June 2013 by bringing the two into its portion of the divided block itself.[873]

Swaziland.

There are no known hydrocarbons in Swaziland.

Tanzania.

LNG Export Trains Possible.

Like Mozambique, Tanzania has also been the site of recent large-scale offshore natural gas discoveries. These new discoveries have boosted its natural gas reserves to over 34 tcf.[874] While there have not yet been any commercial discoveries of **oil** in Tanzania, IOC interest is intense, and there were 16 multinational companies conducting petroleum E&P in Tanzania in 2012 under 23 PSAs.[875] The UK's BG Group, the 60 percent share operator of Blocks 1, 3, and 4, announced in December 2012 its seventh consecutive successful well, making the case that its discoveries in the Jodari Field in Block 1 could anchor Tanzania's first multi-train LNG development.[876] The firm's 40 percent share partner, the UK's Ophir, has targeted first gas in 2018,[877] but given the technical, political, and financial challenges of such a mega-project, slippage in this date appears likely. Ophir announced in April 2013 that it had raised its estimate for the Jodari Field to 4.1 tcf of mean recoverable resources and to 12.6 tcf for blocks 1, 3, and 4.[878] BG drilled the first successful Cretaceous deepwater discovery in May 2013 on Mzia-2, approximately 22 km north of Jodari and also in Block 1.[879] In May 2013, partner Ophir raised its recoverable resource estimate for the Mzia field to 4.5 tcf.[880]

Norway's Statoil, the operator of offshore Block 2, announced five successful wells, including Zafarani, over a 13-month period ending March 2013 totaling an estimated 15-17 tcf in natural gas.[881] Statoil has a 65 percent share in Block 2, with ExxonMobil holding the remaining 35 percent. BG, Ophir, Statoil, and ExxonMobil may join forces to build an LNG plant,[882] with BG planning to present proposed locations for a large LNG terminal to the Tanzanian government in the coming months.[883] The UK's Aminex and Solo Oil have, respectively, 75 and 25 percent shares in the 20 percent onshore, 80 percent offshore Rovuma PSA in southeast Tanzania, where they have successfully drilled two exploratory wells, including Ntoya-1 in the Mtwara Block that is estimated to hold 1.17 tcf of unrisked gas.[884]

Following Statoil's discovery of four more tcf in June 2012 in Block 2, Tanzanian Energy and Minerals Minister Sospeter Muhongo announced in late June that the government planned to introduce new legislation for the gas industry. The Minister stated:

> Deepwater offshore oil and gas exploration is a very expensive undertaking. It is important to ensure this gas and oil exploration momentum continues, and the government will give a bigger push for companies to proceed.[885]

Muhongo announced in July 2012 that royalties on gas production would rise from the current 12.5 percent, with the actual increase to be determined in conjunction with the new legislation.[886] The Minister also announced an increase in signing fees and a review of existing contracts — the latter, he later stressed in September 2012, would only be carried out to avoid mistakes in negotiating **future** deals.[887] Consistent with building momentum, Tanzania announced that it would soon offer nine new southeast blocks adjacent to the blocks where Anadarko and ENI made large gas discoveries offshore Mozambique, in depths from 2,000 to 2,500 meters, possibly in 2013 4Q.[888] Seven more blocks, further off the country's Indian Ocean coast in depths of 3,000 meters, will be auctioned off at a later date.[889]

The UK's Afren has picked up a 74 percent interest in the Tanga Block from Petrodel Resources; the block is onshore and offshore in northeast Tanzania and adjacent to Afren's L17/L18 blocks in southern Kenya.[890] France's Maurel et Prom owns a 60 percent exploration permit to the onshore Bigwa—Rufiji and Mafia Block, where "gas resources [are] to be developed."[891] Australia's Swala Energy has two onshore licenses in the West Rift Valley for which it will carry out 2D seismic surveys in 2013 2Q, Kiloa-Kilombero in east-central Tanzania and Pangani running parallel to its northern border with Kenya.[892] In 2011, Australia's NuEnergy signed joint venture agreements with Tancoal and Tanzacoal for exploration for coal-bed methane in the two firms' coal fields in southern Tanzania near the Malawi border. NuEnergy began scout exploration drilling during 2012 2H.[893]

In February 2013, the government announced that it would soon issue a license to France's Total to explore in Lake Tanganyika, on the border with DRC.[894] In March 2013, Australia's Jacka Resources signed a PSA with the government for the Ruhuhu Block onshore in southwest Tanzania. The Ruhuhu Basin is one of a chain of Karoo geological era basins across southern Africa and abuts the northeast shore of Lake Malawi (Nyasa).[895]

Monetizing Gas Resources Through Domestic Investment.

The Tanzanian government also promoted the expansion of two major projects to monetize its natural gas resources:

1. Songo Songo: Orca currently supplies gas from the Songo Songo Field primarily for power generation to the Tanzania Electric Supply Company, Songas, and 38 industrial customers in the Dar Es Salaam area, including PTCC's cement plant.[896] It plans to increase its gas production capacity from 113 million cf/d to over 200 million cf/d in the coming month. Orca also has an ongoing drilling program to double its production from the Songo Songo Field.[897]

2. Mnazi Bay: 60 percent operator Maurel et Prom and 40 percent partner Wentworth of the UK continue to develop the Mnazi Bay Concession in the Rovuma Basin and drilled a successful well in February 2012.[898] Other than supplying one local power plant, however, this resource had been largely stranded since the field's discovery in 2012. Tanzania's government and these partners, however, are changing this. In October 2012, the government secured a $684 million loan from China Export-Import Bank to build a new 300 MW power plant at Mnazi Bay fed by 75 million cf/d in natural gas. In November 2012, the government officially started construction of the Mnazi Bay to Dar Es Salaam Gas Pipeline Project, with a gas allocation of 80 million cf/d, with an increase to 200 million cf/d as soon as practicable. Finally, Wentworth, with the support of the Tanzanian government, is investigating the construction of the Mnazi Bay Petrochemicals Project to produce ammonia/urea (e.g., for fertilizer) and methanol.[899] In June 2012, the Tanzanian government concluded a deal with China for a $1.225 billion loan from the China Export-Import Bank for the construc-

tion of a gas pipeline from Mtwara to Songo Songo, where it would connect to another pipeline to Dar Es Salaam.[900]

Border Dispute with Malawi over Hydrocarbons.

As noted previously, Tanzania called on Malawi to stop exploring for oil and gas in Lake Malawi (known as Lake Nyasa in Tanzania) until a long-standing border dispute between the two countries can be resolved. Both nations have agreed to have the African Forum mediate their dispute. Malawi claims sovereignty over the entire lake, while Tanzania says it is entitled to 50 percent of it. Malawi awarded two blocks to explore the lake in 2011, while Tanzania awarded one block to explore its claim in 2012.

Togo.

Italian major ENI announced in June 2012 that geological studies and two offshore exploratory wells drilled 17 and 76 km offshore had confirmed the existence of oil in its 100 percent share PSA Blocks 1 (Oti 1) and 2 (Kara 1) in the Dahomey Basin. ENI plans two more wells to determine the quantity and quality of oil. The firm's Togo representative was "optimistic [about commercial quantities of oil] to the extent that [Togo neighbors] Benin and Ghana have substantial oil reserves in the same zone."[901] Four U.S. oil companies had carried out seismic studies and drilling off Togo's coast since the 1960s without finding commercial quantities of oil.[902] ENI's blocks are bordered to the west by the analogous Tano Basin in Ghana, where major discoveries had been made previously,[903] including a find announced by ENI in September 2012.[904]

Tunisia.

Unlike its oil-producing neighbors Algeria and Libya, Tunisia's upstream industry is modest. Tunisia had only 425 million barrels in oil reserves at the end of 2011, equivalent to 15 years' production at the current average of 77,600 barrels of crude per day in 2011. The country no longer exports crude oil, as domestic consumption has risen in recent years,[905] and oil production has declined for most of the last 20 years.[906]

Despite this modest current production, 57 IOCs and local companies conducted exploration and production activities in Tunisia as of 2011 and held 54 exploration licenses for crude oil and natural gas.[907] These large and small IOCs include Spain's Repsol, Australia's Cooper Energy, Canada's Chinook Energy and Sonde Resources, Norway's DNO Resources, Sweden's PA Resources, the UK's Gulfsands, and Vietnam's PetroVietnam.[908]

Areas of hydrocarbon importance include the Gulf of Gabes and the Ghadames Basin in the southern part of the country.[909] The UK's BG Group is the main oil and gas producer in Tunisia and accounted for 13 percent of production in 2009, ahead of NOC ETAP with 12.3 percent and Italy's ENI with 5.6 percent.[910] BG's main source of natural gas was the Miskar Field, in which it held a 100 percent share, but added the Hasdrubal Field in 2009 (50 percent each for BG and ETAP).[911] In 2012, Chinook Energy successfully drilled four horizontal wells in the Sud Ramada field in the Ghadames Basin and plans to drill six more wells there in Bir Ben Tartar Concession in 2013; the firm also plans to undertake a 3D survey of its offshore Cosmos project in 2013, with its first well there in 2014.[912]

Uganda.

Anglo-Irish Tullow Oil first discovered hydrocarbons in Uganda in 2006 with then UK partner Heritage Oil & Gas, whose share it subsequently bought out.[913] In February 2012, Tullow farmed out part of its interests to China's CNOOC and France's Total, with each taking a one-third share in Blocks EA-1, EA-2, and the former EA-3, now known as the Kanywataba and Kingfisher licenses, and each as operator of one of these three blocks.[914]

In contrast to the situation in Ghana, where Tullow has a smooth relationship with the host government, the firm and other IOCs have a more contentious relationship with the Ugandan government. The government claimed in 2010 that Heritage owed over $400 million in taxes because of the sale of its local assets to Tullow, but Heritage claims it owes no taxes; the two sides are in arbitration in London. In December 2012, the Ugandan government reportedly blocked the UK's Ophir Energy from taking over Block 4B when it acquired Dominion Petroleum, claiming it was never notified of the transaction. Subsequently, Canada's Octant Energy announced in March 2012 that it was entering in to a letter of intent to farm-in on Block 4B, but it is uncertain whether this sale of a partial interest will be consummated. At issue is a tax dispute similar to the one between the government and Heritage Oil over its sale of its stake to Tullow. In a separate issue, Tullow filed a case in December 2012 with the World Bank-affiliated International Center for Settlement of Investment Disputes challenging the government's 18 percent value-added tax on imported machinery and supplies.[915]

Yet another contentious issue has been the Uganda government's desire to build a domestic refinery. Uganda President Yoweri Museveni told the Ugandan parliament in December 2012 that one of the common mistakes of African countries was to lose jobs and oil by-products, and increase pipeline transit fees, by exporting crude oil instead of refining it domestically. Some third-party observers agreed with Museveni that a business case could be made for a large-scale refinery in Uganda.[916] President Museveni estimated that Uganda would lose $40 per barrel by exporting crude for refining, in part because crude from Uganda is waxy and the 1,300-km pipeline to Mombasa, Kenya would have to be heated, which adds to the export bill.[917] For their part, partners Tullow, Total, and CNOOC prefer to sell crude on the world market instead of to the refinery and are considering a $5 billion pipeline to the east African coast. In April 2013, the government announced that it and three partners agreed to "start with the refinery size of 30,000 bpd," which implies that the refinery could be enlarged in the future.[918] As of May 2013, the government and the firms were on the cusp of an agreement that would also layout the capacity of the export pipeline, although not include its planned route.[919] One wild card is the construction of a pipeline from Rwanda, through Burundi, to Uganda and the commercial impact that it might have on the potential feasibility of the Uganda Kenya pipeline. (See discussion in the section on Rwanda.)

Tullow's founder and CEO Aldan Heavey told the Ugandan Chamber of Mines & Petroleum in October 2012 it would cost oil companies $12-14 billion to develop the Lake Albertine region and that it might "take 20 years to get that money back . . . if you only have refinery, you have to say how much oil can we sell to the refinery?"[920] To spur greater competition with existing IOCs in Uganda, President Museveni met with Russian President Vladimir Putin in Moscow on December 11, 2012, to discuss bilateral cooperation on energy and mining. Russia's Lukoil has expressed interest in exploration, production, and refining of Uganda's waxy oil, according to the Russian media.[921] The new petroleum law passed by parliament in December 2012 set clear guidelines under which the oil sector is to operate, but there were also concerns about too much discretion given to the Energy Minister (and by implication to Uganda's President) in granting and revoking oil licenses, negotiating PSCs, and in promoting or not promoting transparency in the energy sector.[922]

Uganda has exploration blocks that run from the Sudan border in the north through Lake Albert on the western border with the DRC and south to Lake George. Only five blocks have been licensed;[923] the government is considering a new licensing round with 13 blocks following parliament's passage of a new petroleum law in December 2012.[924] Lake Albert is part of the Albertine Graben, which stretches from the northern end of Lake Albert to the southern end of Lake Tanganyika.[925]

Despite the government's difficulties in working with IOCs, Uganda remains highly attractive to investors because the country's oil discovery rate is an extraordinary 90 percent.[926] By another estimate, only three of 71 wells drilled in the Lake Albert Graben since 2006 have not encountered hydrocarbons. The average cost, including seismic surveys and appraisal drilling to find commercial reserves in Uganda has been about $1 per barrel, compared to an average worldwide cost of $5 to $25 per barrel. Moreover, less than 40 percent of the Lake Albert Graben with potential for oil and natural gas has been explored.[927] This is in only one of five sedimentary basins in Uganda; the others have never been explored.[928]

Estimates of Uganda's oil reserves have varied widely, with Tullow's website indicating one billion barrels and the government of Uganda revising upward reserves to 3.5 billion in September 2012 after appraisal activities on two blocks.[929] The government appointed a panel in July 2012 to work with IOC partners on an overall development plan for the Lake Albertine rift basin, which is rich in biodiversity. Oil production from the basin is estimated at 200,000 bpd 36 months after the development plan is approved,[930] with further increases in daily production possible. First oil production has been repeatedly delayed because of government inaction or disagreements with IOCs;[931] Uganda is expected to begin small-scale oil production in 2014 or 2015.[932] Full-scale commercial production is projected to start after 2017 with full capacity from 120,000-200,000 bpd unlikely to be reached before 2020.[933]

Western Sahara.

IOCs have sought exploration licenses in Western Sahara despite a UN resolution against such deals until the issue of the territory's sovereignty has been resolved. Western Sahara, a desert area bordering the Atlantic Ocean between Mauritania and Morocco that was formerly a colony of Spain, is claimed by Morocco and the Polisario, an independence movement based in Algeria that has proclaimed the Saharawi Arab Democratic Republic (SADR). While information is sketchy, the SADR granted two petroleum exploration licenses to IOCs despite a UN legal notice questioning the legality of such exploration.[934]

Morocco's NOC, ONHYM, reportedly granted France's Total in 2012 a huge block covering 100,926 km² offshore southern Western Sahara under a reconnaissance contract. However, this leasing of acreage is illegal under UN rules.[935] For its part, in 2006 the SADR granted a subsidiary of the UK's Wessex Petroleum and two other UK companies a PSC covering the Bojador onshore block. In 2007, Wessex agreed with Comet Petroleum, the 100 percent holder of the Guetta offshore block, to cross-assignment of interests with the Bojador Block group. Wessex also has a 50 percent interest of the Imlili offshore block east of the Guelta Block, with Tower Resources holding the remaining 50 percent. (Tower purchased Comet in 2008.) Wessex's intention is to:

> commence an initial technical evaluation of the blocks using the existing dataset, and then, subject to the United Nations settlement of the issue of sovereignty of the Western Sahara, to find a substantial farm-in partner.[936]

The SADR granted UK's Ophir Energy a 50 percent operated interest in four blocks with a gross area of 74,327 km².[937] The UK's Premier Oil purchased a 35 percent share of the rights of the U.S. firm Fusion Oil & Gas in 2003 that had been granted by the SADR.[938]

Zambia.

Zambia has no proven oil or gas reserves, but the presence of hydrocarbons has been confirmed in Chavuma and Zambezi, two districts in the Northwestern

Province.[939] Zambian Minister of Mines and Mineral Development, Maxwell M. B. Mwale, indicated in 2010 that:

> Exploration for oil and gas [in Zambia] is being conducted based on the understanding that oil and gas has been discovered in countries with . . . the same geology. Uganda discovered oil along the East African Rift System, which extends to . . . Lake Tanganyika— Mweru Wantipa Area, Luangwa, Luano–Lukusashi, Kafue and Zambezi Valleys. Botswana, on the other hand, discovered coal-bed-methane In the Okavango Basin, which stretches into the northwestern and western provinces of Zambia.[940]

In 2009, the Zambia government launched a first bid round for 23 blocks and awarded 11 blocks. It awarded seven blocks to local Zambian companies Majetu (Blocks 3 and 11), Barotse Petroleum (20 and 21), and, Chat Milling (5, 16, and 19), and four to foreign companies, including the U.S. firm Glint Energy (Block 2, next to the Angolan border to the north), the UK's Petrodel Resources (Block 17, next to the Angolan border to the west),[941] the UK's GB Petroleum (Block 22, in southwest Zambia), and Canada's Exile Resources (Block 26, in the northeast, close to the Malawi border). (Exile was subsequently purchased by Nigeria's Oando Petroleum.)

Zambia's second round of bidding in 2010 did not attract any other foreign investors, and the government held off on awarding any new blocks to local companies.[942] A third round of bidding was held in 2011, with the U.S. firm Frontier Resources winning Block 34, Exile being awarded a second block, and a Zambian company winning a third block.[943] South Africa's Rapid African Energy was awarded Block 31 in northern Zambia on the borders with the DRC, including Lake Mweru, and Tanzania, including Lake Tanganyika.[944] The Zambian government plans a fourth round of bidding in 2013 and plans to offer 26 new blocks and six blocks that were offered in earlier rounds for which there was no suitable bidder.[945]

In February 2013, Zambian Mines Minister Yamfwa Mukanga warned that companies holding exploration licenses in Zambia needed to move forward with their work or risk losing their licenses.[946] One company that claims to have been active is Frontier Resources, whose Block 34 is 150 km southwest of the capital, Lusaka. Frontier believes its block may form part of the southwestern extension of the Western Rift Valley branch of the East African Rift System,[947] has been acquiring 2D data near the block shot by an earlier operator, and is reviewing the acquisition of new aerogravity/aeromagnetic data.[948]

Zimbabwe.

Zimbabwe has no proven oil or gas reserves, but has potential to become a producer of coal-bed methane gas.[949] Minister of Mines and Mining Development Obert Mpofu said in 2011 that Zimbabwe had forged a deal with a[n undisclosed] foreign partner to explore and develop the "massive oil reserves" that the government believes are located in the Zambezi Valley. The Minister also stated that "There has been a lot of exploration along the Zambezi Valley for oil. We have another option to extract oil from coal, but . . . will only consider that when all other avenues have been exhausted." The Minister acknowledged that exploration was still at the initial stage that involves the study of structural geology at the reservoir scale and sedimentary basin analysis.[950] Another source claimed that Zimbabwe contained the largest reserves of methane gas in Africa,[951] which could be true given Zimbabwe's extensive coal beds. Consistent with this, Australia's Sunbird Energy has eight permits for exploration in the Kalahari Karoo Basin in northern Botswana, approximately 100 km south of the Wankie coalfield in Zimbabwe.

Firms such as Sunbird may be choosing to invest in coal beds in neighboring countries such as Botswana, South Africa, and Zambia because of the current Zimbabwe government of Robert Mugabe's hostile attitude toward foreign investment.

ENDNOTES - APPENDIX I

1. "Algeria—Country Analysis Brief," U.S. Energy Information Agency (EIA), March 8, 2012, lists Libya as ahead of Algeria in crude oil production, whereas Mowafa Taib, "The Mineral Industry of Algeria," *U.S. Geological Survey Mineral Yearbook-2010*, Vol. III, New Cumberland, PA: U.S. Geological Survey, 2010, p. 2.1, puts Algeria in third place.

2. Mowafa Taib, "The Mineral Industry of Algeria," pg. 2.1.

3. "Algeria—Country Analysis Brief," EIA.

4. *Ibid.*; "International/Frontier Producing—Algeria," available from *www.anadarko.com/operations*.

5. "Key Operating Areas," available from *www.cnoocltd.com/encnoocltd*.

6. "CNPC in Algeria," available from *www.cnpc.com.cn/en/cnpcworldwide/algeria*.

7. See *www.gazprom.com/about/production/projects/deposits/algeria*.

8. "Algeria Hassi Bir Rekaiz Project," September 30, 2012, available from *www.pttep.com/en*.

9. "ConocoPhillips to Sell Off Algeria Unit," *Petroleum Africa*, December 19, 2012.

10. Lahcen Achy, "Algeria Needs More than Hydrocarbon Law Amendments," *Al-Hayat*, January 22, 2013.

11. "Algeria Assembly Approves Amendments to Energy Law," *Reuters*, January 21, 2013.

12. "Algeria Plans for Bid Round," *Petroleum Africa*, November 8, 2012.

13. Achy, "Algeria Needs More than Hydrocarbon Law Amendments."

14. "Algeria—Output Slides on Joint Ventures," April 12, 2012, available from *www.africaintelligence.com*.

15. Achy, "Algeria Needs More than Hydrocarbon Law Amendments."

16. John Mitchell *et al.*, "Resource Depletion, Dependence, and Development: Algeria," London, UK: Chatham House, November 2008, p. 18.

17. Achy, "Algeria Needs More than Hydrocarbon Law Amendments."

18. "Algeria—Country Analysis Brief," EIA.

19. Taib, "The Mineral Industry of Algeria," p. 2.4.

20. "Repsol Makes Illizi Discovery," *Petroleum Africa*, April 9, 2013.

21. "Repsol in Algeria," available from *www.repsol.com/es en/corporacion/conocer-repsol/quienes-somos/presencia-global/argelia*.

22. "ENI Announces Start-Up of Gas Production from MLE Field in Algeria," February 4, 2013, available from *www.eni.com/ en IT/media/press-releases/press-release.page.*

23. "Algeria," available *www.statoil.com/AnnualReport2011/ en/Pages/frontpage.aspx*; and "BP in Algeria," available from *www. bp.com.*

24. "Production: Algeria, Gassi El Agreb Redevelopment Project," available from *www.hess.com.*

25. "El Merk Delivers for Anadarko," *Petroleum Africa*, May 8, 2013.

26. *Ibid.*; and "Algeria," Vietnam National Oil and Gas Group, available from *english.pvn.vn*. Note: PetroVietnam's website indicates the NOC has an exploration and appraisal contract for Block 433A and 416B in Algeria.

27. "Algeria—Output Slides on Joint Ventures," April 12, 2012, available from *www.africaintelligence.com.*

28. Paul Williams, "PTTEP, Sonatrach, and CNOOC Discover Crude Oil in Algeria," September 21, 2012, available from *www. offshore-publication.com.*

29. "CNPC in Algeria."

30. "Algeria Sets Aside $80 Billion for Energy Investments," November 5, 2012, available from *www.ventures-africa.com.*

31. "Doing Business in Algeria: 2012 Commercial Guide for U.S. Companies," Washington, DC: U.S. Department of State, U.S. Foreign Commercial Service (USFCS).

32. "Upstream Africa Acreage—Algeria," *Total Factbook 2011*, Washington, DC: Central Intelligence Agency, 2011.

33. "Algeria—Country Analysis Brief," EIA.

34. "245-South Block (Algeria)," available from *www. rosneft.com.*

35. "Russian Appraise Algerian Discoveries," *Africa Oil and Gas Report*, May 2012, p. 16.

36. EAI Short-term Energy Outlook (STEO), as provided by USAFRICOM source on April 25, 2013.

37. "Angola Background," Washington, DC: EIA, January 8, 2013.

38. "Angola," available from *www.mbendi.com/land/af/an/p0005.htm*.

39. *Ibid.*; and "Total Leaves Them All in the Dust," *Africa Oil and Gas Report*, September 3, 2012.

40. "Angola Background," EIA.

41. "Angola LNG Up and Running," *Petroleum Africa*.

42. "Angola Background," EIA.

43. "Angola Oil—South West Africa's Next Crown Jewel," May 16, 2011, available from *mergersandacquisitionreviewcom. blogspot.com/2011/05/angola-west-africas-next-crown-jewel.html*.

44. *Ibid.*

45. "Oil, Gas and Security in Angola," available from *thinksecurityafrica.org/oilgas/oil-and-gas-in-Angola*.

46. "Angola Oil—South West Africa's Next Crown Jewel."

47. "Another Angola Discovery for ENI," *Petroleum Africa*.

48. "Repsol in Angola."

49. Oswald Clint and Rob West, "Angola's Pre-Salt Provinces—the Next Brazil?" Oxford, UK, Oxford Energy Forum, November 2012.

50. "Research and Markets: Offshore Drilling Industry in Middle East and Africa to 2016," March 25, 2013, available from *Businesswire.com*.

51. "Can Chevron Close the Gap on Total in Angola?" *Africa Oil & Gas Report*, February 25, 2013.

52. "Exploration and Production Operations in Africa— Angola," *www.total.com*.

53. "Can Chevron Close the Gap on Total in Angola?"

54. "Angola," available from *www.statoil.com/AnnualReport 2011/en/Pages/frontpage.aspx*.

55. "Profile: Angola," available from *www.exxonmobil.com*.

56. "BP in Angola," available from *www.bp.com*; "Projects and Achievements—Girassol," available from *www.total.com*.

57. "Angola Oil—South West Africa's Next Crown Jewel."

58. "Mafumeira Sul Gets Go Ahead from Chevron," *Petroleum Africa*, February 7, 2013.

59. "Angola Fact Sheet," April 2012, available from *www.chevron.com*.

60. "Angola Oil—South West Africa's Next Crown Jewel"; and "Angola Background," EIA.

61. "Our Business," available from *www.chinasonangol.com/ our business.html*.

62. "Angola," available from *www.marathonoil.com/Global Operations/Angola/Operations*.

63. See *www.conocophillips.com/zmag/SMID 392 FactSheet-OtherInternational.html*.

64. "Maersk Sees Success with Angola's Block 16 Well," *Petroleum Africa*, September 25, 2012.

65. "West Africa," available from *www.cobaltintl.com*.

66. "Cobalt Pushes Pre-Salt Program in West Africa," *Petroleum Africa*, February 28, 2013.

67. "Cobalt Explores Pre-Salt Potential," *African Energy*, May 2, 2013.

68. "Angola Background," Washington, DC: U.S. Energy Information Administration, January 8, 2013.

69. "Angola Prepares Onshore Licensing Round," *Petroleum Africa*, April 5, 2013.

70. "Angola Lines Up Onshore Bid Round," *AfrOil*, April 9, 2013.

71. "Angola/ROC Sign Agreement for Joint Exploration," *Petroleum Africa*, March 20, 2012.

72. "Accord Paves Way for Chevron Angola/ROC Development," *Petroleum Africa*, July 31, 2012.

73. "Angola Background," EIA.

74. "Benin," available from *www.mbendi.com/land/af/be/p0005.htm*.

75. "Sillinger Looks Forward to Unlocking Benin's Potential," *Petroleum Africa*, October 12, 2011.

76. "Benin: des gisements de petrole ont ete decouverts au large des cotes du Benin" ("Benin : Oil Deposits Discovered Along Coast of Benin"), February 3, 2009, available from *westafrica.sme-toolkit.org*.

77. "Benin."

78. "Oil, Gas, and Security in Benin," available from *thinksecurityafrica.org/oilgas/oil-and-gas-in-Benin*.

79. "Benin Contract Extension," available from *www.sapetro.com*; and John Ankromah, "Seme Field Returns to Production in 2014," *Africa Oil & Gas Report*, March 25, 2013.

80. "Oil, Gas, and Security in Benin," available from *think securityafrica.org/oilgas/oil-and-gas-in-Benin*.

81. Century International Oil & Gas Corporation, "West Africa Offshore Benin," *NYTimes.com*, June 2012.

82. See *www.poloresources.com/Investments Signet.htm*.

83. "Petrobras Enters Benin," *Petroleum Africa*, February 24, 2011; and see *www.petrobras.com*.

84. "Deepwater Blocks 5 & 6 Benin Offshore Farm in Opportunity," available from *www.simco-pet.com*.

85. "Oil and Gas Availability in Botswana," available from *thinksecurityafrica.org/oilgas/oil-and-gas-in-Botswana*.

86. "Sasol Announces Joint Venture Participation with Origin Energy in Coal Bed Methane Exploration," November 1, 2011, available from *www.sasol.com*; and "Origin Energy Shareholder Review 2012," available from *reports.originenergy.com.au/2012/shareholder/*.

87. Josh Lewis, "Duo to Explore Botswana CBM," November 2, 2011, available from *www.upstreamonline.com*.

88. See *www.sunbirdenergy.com.au/Botswana/kasane-project.html*.

89. Iain Esau, "Tlou in Botswana CBM Push," February 28, 2013.

90. "Botswana CBM," available from *www.magnumgpl.com*.

91. "Upstream update : Zambia, Malawi, and Botswana," *African Energy*, May 16, 2013.

92. "Botswana Petroleum," available from *www.magnumgpl.com*.

93. Desire Nimubona, "Le Burundi sur le point de produire du petrole" ("Burundi About to Produce Oil"), September 26, 2012, available from *www.isanganiro.org*.

94. "Burundi," available from *www.poloresources.com/Investments Signet.htm*.

95. Nimubona, "Le Burundi sur le point de produire du petrole" ("Burundi About to Produce Oil").

96. "Burundi: Lake Tanganyika Exploration Eyed," *Oil & Gas Journal*, February 3, 2011.

97. "East Africa Rift System: Lake Tanganyika, Burundi Blocks B and D Farm-in Opportunity," Surestream Petroleum, April 30, 2012.

98. "Oil, Gas and Security in Cameroon," available from *www.thinksecurityafrica.org/oilgas/oil-and-gas-in-Cameroon*; and "Cameroon Oil and Gas Profile," available from *www.abarrelfull.wikidot.com*.

99. "Cameroon Oil and Gas Exploration in 2011 and 2012," June 20, 2011, available from *mergersandacquisitionreviewcom.blogspot.com/2011/06/cameroon-oil-and-gas-exploration-in.html*.

100. Jean Jacques Nguimbous-Kouoh *et al.*, "Structural Interpretation of the Mamfe Sedimentary Basin of Southwestern Cameroon along the Manyu River Using Audiomagnetotellurics Survey," *ISRN Geophysics*, Vol. 2012, Article ID 413042, 2012.

101. "Cameroon," available from *www.mbendi.com/land/af/ca/p0005.htm*.

102. "Cameroon," available from *www.bowleven.com*.

103. "Oil, Gas and Security in Cameroon," available from *www.thinksecurityafrica.org/oilgas/oil-and-gas-in-Cameroon*.

104. "Regions Petroliferes en Afrique" ("Petroleum Regions in Africa"), available from *fr.wikipedia.org*.

105. "Cameroon Hosting Bid Round," *Petroleum Africa*, March 8, 2013.

106. "Cameroon," available from *www.perenco.com/operations/ Africa/Cameroon*.

107. "Cameroun: de nouvelles reserves de petrole et gaz naturel," (Cameroon: New Oil and Gas Reserves), October 16, 2012, available from *www.legriot.info*; "Cameroon Plans for Crude Boost," *Petroleum Africa*, January 26, 2012.

108. "Cameroon," available from *www.perenco.com/operations/ Africa/Cameroon*.

109. "Cameroon—License Areas," available from *www. addaxpetroleum.com/operations/cameroon*.

110. "Sinopec Enters Cameroon," *Petroleum Africa*, November 8, 2011.

111. "Bowleven Sees Even More Pay at Sapele," *Petroleum Africa*, January 11, 2011.

112. "Sapele-3 Hits for Bowleven in Cameroon," *Petroleum Africa*, October 17, 2011; "Bowleven Nears TD Offshore Cameroon," *Petroleum Africa*, February 1, 2013; "Bowleven has Intra Isongo Discovery Offshore Cameroon," *Oil & Gas Journal*, February 2013.

113. "Cameroon: Oil and Gas Exploration in 2011 and 2012," June 20, 2011, available from *mergersandacquisitionreviewcom. blogspot.com/2011/06/cameroon-oil-and-gas-exploration-in.html*.

114. "Cameroon," available from *www.bowleven.com*.

115. "Bowleven va investir $900 millions au Cameroun" ("Bowleven Will Invest $900 Million in Cameroon"), March 30, 2012, available from *AfricaTime.com*.

248

116. "Cameroon," available from *www.afexglobal.com/ operations/cameroon.*

117. "Cameroon," Vietnam National Oil and Gas Group, available from *english.pvn.vn.*

118. Toyin Akinosho, "Kosmos Throws $75 Million in the Hole," *Africa Oil & Gas Report,* May 21, 2013.

119. "Cameroon," available from *www.kosmosenergy.com/ operations-cameroon.*

120. "Cameroon: Kosmos Energy Signs PSC for Fako Block, Onshore Cameroon," January 15, 2012, available from *www. energy-pedia.com/news/cameroon/new-148724.*

121. "Republic of the Congo," available from *www.murphyoil-corp.com.*

122. Tess Stynes, "Murphy Oil Offshore Congo Well Proves Unsuccessful," *The Wall Street Journal,* November 21, 2012.

123. "Cameroon: Ntem," available from *www.sterlingener-gyuk.com.*

124. "West Africa," available from *www.nobleenergyinc.com.*

125. "Cameroon: Oil and Gas Exploration in 2011 and 2012," June 20, 2011, available from *mergersandacquisitionreviewcom. blogspot.com/2011/06/cameroon-oil-and-gas-exploration-in.html.*

126. "NBL Noble Energy," pdf presentation at Noble Energy Analyst Conference, December 6, 2012.

127. "No Trema Oil for Noble Energy (Cameroon)," October 18, 2012, available from *www.offshoreenergytoday.com.*

128. "Africa—Cameroon," available from *www.dana-petroleum.com/Our-businesses/Africa/Overview.*

129. "Cameroon—Expanding into Cameroon," available from *www.madisonpetro.com.*

130. "Nigeria/Cameroon to Jointly Explore Bakassi Region," *Petroleum Africa*, March 14, 2011.

131. "Cameroun: La Chine s'investit dans le petrole et le gaz naturel" ("Cameroon: China Invests in Oil and Natural Gas"), April 5, 2009, available from *accypresse.over-blog.com/article-29907951.html*.

132. Emmanuel Tumanjong, "Floods Halt Exploration of Chinese Company in N Cameroon," January 11, 2013, available from *www.rigzone.com*.

133. "Cameroon," available from *www.bowleven.com*.

134. Philippe Nsoa, "Cameroun: Le gaz naturel cherche sa voie" ("Cameroon: Natural Gas Finds its Path"), September 8, 2011, available from *www.ajafe.net*.

135. "Cameroun LNG," available from *www.gdfsuez.com/wp-content/uploads/2012/05/cameroun-uk.pdf*.

136. Achille Mbog Pibasso, "D'importantes reserves de petrole et de gaz" ("Important Oil and Gas Reserves"), *Les Afriques*, March 7-13, 2013.

137. "Cameroon," available from *www.perenco.com/operations/Africa/Cameroon*.

138. "VOG Steps Toward Commercialization at Logbaba," *Petroleum Africa*, June 14, 2011.

139. "Cameroon: Recent History," available from *www.victoriaoilandgas.com*.

140. "Country Commercial Guide — Cameroon," Washington, DC: Department of State, USFCS.

141. "Brazil's Petrobras to Support Cape Verde in Ultra-deepwater Oil and Gas Prospecting," July 5, 2010.

142. "Cameroon and Cape Verde—2010 [Advance Release]" *U.S. Geological Survey Minerals Yearbook 2010*, New Cumberland, PA: U.S. Geological Survey, 2010.

143. "Sonagol Wants Refinery and Oil and Gas Wells in Cape Verde," June 5, 2006, available from *www.macauhub.com*.

144. "Petrole: Les Chinois reprennent la recherche de petrole en Centrafrique" ("Oil : The Chinese Restart Oil Exploration in the Central African Republic"), February 1, 2012, available from *commodesk.com*.

145. David-Christian Vonto, "Réflexion Sur La Possibilité d'Implantation d'une Raffinerie en RCA" ("Reflections on the Building of a Refinery in the CAR"), 2006, available from *www. sozowala.com*.

146. "Central African Republic," *World Factbook*, available from *www.cia.gov*.

147. "Ousted Central African Republic leader Bozize seeks exile in W. African Nation of Benin," *The Washington Post*, March 29, 2013.

148. "Self-Proclaimed President Michel Djotodia Elected Interim Central African Republic President," April 13, 2013, available from *www.breakingnews.com*.

149. See *www.mbendi.com*, quoting 2012 BP Statistical Energy Survey.

150. "Chad Oil History," Geointelligence Network, available from *www.geosint.com*.

151. See "Chad," *www.mbendi.com/land/af/ch/p0005.htm*.

152. "Oil, Gas, and Security in Chad," March 5, 2013, available from *thinksecurityafrica.org/countries/security-in-chad/*.

153. "Chapter 1: Doing Business in Chad," *2012 Country Commercial Guide for Chad*, Washington, DC: Department of State, U.S. Foreign Commercial Service, 2012.

154. "Chad Fact Sheet," April 2012, available from *www.chevron.com*.

155. "Profile: Chad/Cameroon—Doba Basin," available from *www.exxonmobil.com*.

156. "The Mineral Industry of Chad," Chad—2011 [Advance Release], *U.S. Geological Survey Minerals Yearbook 2011*, p. 9.1.

157. "Du petrole chinois pour l'oleoduc Tchad-Cameroun" ("Chinese Oil for the Chad-Cameroon Pipeline"), Agence Ecofin, February 16, 2012.

158. "Griffiths to Connect Production to Chad/Cameroon Pipeline," *Petroleum Africa*, January 31, 2013.

159. "Griffiths Energy Provides Development and Exploration Update," Griffiths Energy International press release, January 28, 2013.

160. "Griffiths Progresses Work in Chad," *Petroleum Africa*, March 26, 2013.

161. "Glencore Picks Up Stakes in Griffiths' Chad Blocks," *Petroleum Africa*, September 26, 2012.

162. "Griffiths Energy revives IPO plans," *African Energy*, May 16, 2013.

163. Carrie Tait and Jacquie MacNish, "Police Seize Griffiths Shares After Search of Law Office," *The Globe and Mail*, February 22, 2103.

164. "ERHC Signs EEA with Chad," *Petroleum Africa*, July 19, 2012.

165. "ERHC's Oil and Gas Exploration Interests in Chad," available from *www.erhc.com*.

166. "Simba Energy in Chad," available from *www.simbaenergy.ca/projects/chad.aspx*.

167. "Tchad: Simba Energy dans l'aventure petroliere" ("Chad: Simba Energy in Petroleum Adventure"), *Journal du Tchad*, October 22, 2012.

168. "Simba Takes on Three Chadian Blocks," *Petroleum Africa*, September 25, 2012.

169. As of March 2013, Taiwan authorities maintained diplomatic relations with only four countries in Africa: Burkina Faso, the Gambia, Swaziland, and São Tomé and Principe.

170. "Upstream Operations," available from *www.cpc.com.tw*.

171. "CNPC in Chad," available from *www.cnpc.com/cn/en/cnpcworldwide/chad*.

172. "The Mineral Industry of Chad," Chad—2011 [Advance Release], *U.S. Geological Survey Minerals Yearbook 2011*, p. 9.2.

173. "China, Chad Re-Establish Diplomatic Relations, Upsetting Taiwan," *Voice of America*, October 31, 2009.

174. George Obulutsa, "Comoros Awards First Oil Exploration License," *Reuters*, March 28, 2012; see *www.reuters.com/article/2012/03/28/ozabs-comoros-exploration-idAFJOE82R02X20120328*; and *www.theeastafrican.co.ke/business/Kenyan+company+wins+Comoros+oil+exploration+dal+/-/2560/1386360/-/o1gm0r/-/index.html*.

175. "Comores: Elaboration de la legislation petroliere" ("Comores: Elaboration of Oil Law"), September 21, 2012, available from *www.comores-actualites.com*.

176. "International Arbitration Launched Against Comoros Government," *Petroleum Africa*, November 27, 2012.

177. "Comoros Comments on E&P Agreement," *Petroleum Africa*, October 12, 2012.

178. "Cove Energy Directors Look to Replicate Success in Mozambique," *Reuters*, March 10, 2013.

179. "Former Cove Execs Form New Company," *Petroleum Africa*, March 12, 2013.

180. Comores/Prospection petroliere: marcher comme sur des oeufs" ("Comoros: Oil Exploration Like Walking on Eggs"), October 17, 2012, available from *www.comores-actualites.com.*

181. "Du Protocole avec GTX a la legislation sur le petrole" ("From the GTX Agreement to Oil Legislation"), October 17, 2012, available from *www.comores-actualites.com.*

182. "Core Assets—Comoros," available from *www.avanapetroleum.com/islands.*

183. "Congo," available from *www.mbendi.com/land/af/co/p0005.htm.*

184. The figure for the United States has likely fallen since 2010, however, because the recent rapid increase in U.S. domestic shale oil production has displaced imports.

185. "Upstream Africa Acreage," *Total Factbook 2011.*

186. "ROC's Moho Bilondo Nord Requires Big Investment," *Petroleum Africa,* July 26, 2011.

187. "ROC's Moho Projects Get Greenlight," *Petroleum Africa,* March 25, 2013.

188. See *total.com/fr/energies-savoir-faire/petrole-gaz/exploration-production.*

189. See also figures for Republic of Congo in "Exploration & Production—Activities in the World," available from *www.eni.com.*

190. "Republic of the Congo," available from *www.murphyoilcorp.com/Global-Operations/Atlantic-Margin/#!republic-of-congo.*

191. "Congo (Rep): ExxonMobil farming out Mer Tres Profonde Sud Block, Offshore Republic of Congo," August 4, 2010, available from *www.energy-pedia.com.*

192. "Congo," available from *www.perenco.com/operations/ africa/congo.html*.

193. "Congo (Brazzaville)," available from *www.afren.com/ operations/congo brazzaville/*.

194. "Congo (Brazzaville)," available from *www.maurelet prom.fr*.

195. "Congo," available from *www.socointernational.co.uk/ congo*.

196. "Regions Petroliferes en Afrique" ("Petroleum Regions in Africa"), available from *fr.wikipedia.org*.

197. "Congo," Vietnam National Oil and Gas Group, available from *english.pvn.vn*. Note: PetroVietnam's website indicates the NOC has the marine XI Block contract.

198. "Other Regions in Africa," available from *www.cnoocltd. com*; "Republic of Congo (Brazzaville)," available from *www. paresources.se*; "Svenska Petroleum Exploration Announces the Ivory Coast and Republic of Congo Exploration Farm-Ins," company press release, May 25, 2009; "Congo (Brazzaville)," available from *www.tullowoil.com*; "Upstream Congo," available from *www. vitol.com/vitol-in-west-africa.html*.

199. "CNOOC Congo Hires Drillship 'Jasper Explorer'," January 24, 2013, available from *www.worldmaritimenews.com*; and Ed Reed, "China Lines Up Role in Africa's Rise," *AfrOil*, April 2, 2103.

200. "Congo (Brazzaville) Country Analysis Brief," EIA, December 14, 2011.

201. "ENI to Start Congo Pilot Oil Sands Project in 2012," *Reuters*, October 6, 2011.

202. "Sustainability—The Tar Sands Project," available from *www.eni.com/en IT/world-eni/index.shtml?selectCountry*.

203. "Congo (Brazzaville) Country Analysis Brief," EIA.

204. "Access to Energy and Gas Flaring Projects," available from *www.eni.com/en IT/world-eni/index.shtml?selectCountry*.

205. "Central African Republic, Cote d'Ivoire, and Togo," *U.S. Geological Survey Minerals Yearbook—2010* [Advance Release], p. 8.4.

206. "Oil, Gas and Security in Cote d'Ivoire," available from *thinksecurityafrica.org/oilgas/oil-and-gas-in-ivory-coast/*; and "Oil and Petroleum," (Cote d'Ivoire), available from *en.wikipedi.org/wiki/Energy in Cote d'Ivoire*.

207. "Canadian Natural Resources—Baobab," available from *www.cnrl.com*; and "Cote d'Ivoire—Oil and Gas Discoveries and Status of Exploration Activity in 2011 and 2012," available from *mergersandacquisitionreviewcom.blogspot.com/2011/05/cote-divoire-oil-and-gas-discoveries.html*.

208. "Afren PLC : Operations : Cote d'Ivoire," Blocks CI-01 and CI-11, available from *www.afren.com*.

209. "Svenska Petroleum Exploration Wildcat Finds Gas offshore Cote d'Ivoire," press release, *Svenska*, September 28, 2010.

210. "Vanco Hits Light Oil off Cote d'Ivoire," *Petroleum Africa*, December 8, 2011.

211. "Lukoil Eyes New Ghanaian Drilling," *AfrOil*, April 2, 2013.

212. "Cote d'Ivoire: Lukoil Sign Deal to Seek Oil Offshore Ivory Coast," October 2, 2012, available from *www.energy-pedia.com*.

213. "Lukoil Plans Cote d'Ivoire Spend," *Petroleum Africa*, February 27, 2013.

214. "Paon-1X Exploration Well Discovers Oil in Cote d'Ivoire," news release, *Tullowoil.com*, June 7, 2012.

215. "Cote d'Ivoire—CI-202," available from *www.rialtoenergy.com/*. and "Rialto Energy is Certain of 40 million 'Scuf'," *Africa Oil and Gas Report*, September 4, 2012.

216. "Ivoire 1X Boosts the West African Transform Margin," *Oil & Gas Report*, May 2, 2013.

217. Note: Yam's Petroleum is owned by Lebanese-Ivorian architect Pierre Fakhoury, who reportedly was given five exploration blocks as payment for his work on state infrastructure projects—raising questions about the transparency of how the government awards blocks, at least to domestic investors. "Cote d'Ivoire: Oil Discovery for Total," *African Energy*, May 2, 2012.

218. "Cote d'Ivoire: Total Acquires an Interest in CI-100 Deepwater Exploration License," October 22, 2010, available from *www.total.com*.

219. "Ghana-Cote d'Ivoire: New Find, Old Maritime Dispute," *OilPrice.com*, May 10, 2013.

220. "Total Acquires Interests in Three New Ultra-Deepwater Exploration Licenses," February 17, 2012, available from *www.total.com*.

221. "Cote d'Ivoire," available from *www.africanpetroleum.com.au*.

222. "Our Operations in Cote d'Ivoire," available from *www.genelenergy.com*.

223. "DRC," available from *www.mbendi.com/land/af/zr/p0005.htm*.

224. "Oil, Gas, and Security in the DRC," available from *thinksecurityafrica.org/oilgas/oil-and-gas-in-the-democratic-republic-of-congo/*.

225. "Operations—DR Congo," available from *www.heritageoilplc.com/*.

226. "News over Ownership of Congolese Oil Blocks Raises Further Corruption Concerns," June 29, 2012, available from *www.global witness.org*.

227. "Total Farms-in to DRC's Block III," *Petroleum Africa*, March 4, 2011; "Democratic Republic of the Congo," *Total Factbook*

2011; "Democratic Republic of the Congo (DRC)," available from *www.sacoilholdings.com/operations/drc*.

228. "PetroAfrican Resources Presents its Site Survey Report of Block 4 in the Albert Graben Rift System, Democratic Republic of the Congo," available from *www.petroafricanresources.com*.

229. "DRC," available from *www.socointernational.co.uk/drc*.

230. See *en.wikipedia.org/wiki/SOCO International*.

231. "DR Congo - EC launches environmental study on Albertine Rift," *African Energy*, May 16, 2013.

232. "DRC to Adopt New Oil Law," *Petroleum Africa*, February 1, 2013.

233. "Alberta Oilsands Heads to DRC," *Petroleum Africa*.

234. "ROC and DRC Programs Updated," *Petroleum Africa*, November 2, 2011.

235. Eduard Gismatllin, "Soco to Expand Vietnam Field Output 27%, Allowing Dividend," Bloomberg, March 11, 2013.

236. "Lotshi Block," available from *www.energulf.com/assets/democratic-republic-of-congo/*.

237. "EnerGulf Prepares for DRC Drilling Program," *Petroleum Africa*, April 6, 2011.

238. "Drilling in Lotshi in 2013 — EnerGulf Resources," available from *www.stockhouse.com/bullboards*.

239. "DRC: Lotshi Resource Report In," *Petroleum Africa*, May 12, 2011.

240. "Inpex Pick-up DRC's Ngani Block," *Petroleum Africa*, August 26, 2010; "Offshore D.R. Congo Block," available from *www.inpex.co.jp/english/business/africa.html*.

241. "Resolution of DRC/Angola Dispute on the Way," *Petroleum Africa*, September 26, 2012.

242. "DRC Negotiates with IOCs to Boost Output," *Petroleum Africa*, July 21, 2011.

243. "Le petrole et le gaz de RDC interessent Chevron et Petrobras" ("DRC Oil and Gas Interest Chevron and Petrobras"), Agence Ecofin, July 25, 2011.

244. "Djibouti," available from *www.oysteroil.com*.

245. "Oyster Oil Secures Djibouti Acreage," *Petroleum Africa*, September 20, 2011.

246. "Oyster Awarded Exploration Blocks Off Djibouti," September 20, 2011, available from *somalilandpress.com*.

247. "Promising Results in Preliminary Oil and Gas Search in Djibouti," July 20, 2012, available from *www.sabahionline.com*.

248. Another source indicated that average daily oil production in 2011 was somewhat higher: 735,000 bpd. See *www.mbendi. com/land/af/eg/p0005.htm* quoting 2012 BP figures.

249. "Egypt—Country Analysis Brief," Washington, DC: U.S. Energy Information Administration, July 2012.

250. "Egypt Knocks Out 17 discoveries in Q1," *Petroleum Africa*, May 7, 2013.

251. "Egypt—Country Analysis Brief," EIA.

252. "Egypt—2010" [Advance Release], *U.S. Geological Survey Minerals Yearbook—2010*, New Cumberland, PA: U.S. Geological Survey, p. 13.5.

253. "Egypt—Country Analysis Brief," EIA.

254. "Egypt—2010" [Advance Release], *U.S. Geological Survey Minerals Yearbook—2010*, p. 13.5.

255. "Repsol in Egypt," available from *www.repsol.com/es_en/ corporacion/conocer-repsol/quienes-somos/presencia-global/egipto.aspx*.

256. "Egypt—Country Analysis Brief," EIA.

257. "Egypt—Country Commercial Guide—2012," U.S. Foreign Commercial Service.

258. *Ibid.*

259. "EGAS Announces Block Winners," *Petroleum Africa*, April 18, 2013.

260. "Egypt to Ask for Bigger Production Share," *Petroleum Africa*, May 1, 2013.

261. "Apache plans 270 wells in Egypt," March 5, 2013, available from *www.upi.com*.

262. "Apache has Big Drilling Plans for Egypt," *Petroleum Africa*, March 14, 2013.

263. "Egypt," available from *www.apachecorp.com/Operations/ Egypt/index.aspx*.

264. "Apache Hits Triple in the Western Desert," *Petroleum Africa*, May 8, 2013.

265. "JVs/PSCs/Alliances," available from *www.oil-india*.

266. "BP in Egypt," available from *www.bp.com*.

267. "Egypt—2010 [Advance Release], *U.S. Geological Survey Minerals Yearbook—2010*, p. 13.6.

268. "BP in Egypt," available from *www.bp.com*.

269. "ENI: New Oil Discovery In Western Desert of Egypt," ENI press release, February 6, 2013.

270. "Egypt," available from *www.bg-group.com*.

271. "Royal Dutch Shell Takes Final Investment Decision on Alam El Shawish West Gas Development Project," press release, January 22, 2012, available from *www.shell.com*.

272. "Dana Grows Fatter in Egypt," *Africa Oil & Gas Report*, March 14, 2013.

273. See *en.wikipedia.org/wiki/Energy in Egypt*.

274. "Equatorial Guinea," available from *www.mbendi.com/land/af/eq/p0005.htm*; "Equatorial Guinea," Washington, DC: U.S. Energy Information Administration, available from *www.eia.gov/countries/country-data*.

275. "Oil, Gas and Security in Equatorial Guinea," available from *thinksecurityafrica.org/oilgas/oil-and-gas-in-equatorial-guinea/*.

276. "ExxonMobil Starts Production Offshore Equatorial Guinea," August 5, 2003, available from *www.gasandoil.com/news/africa/5a35ec3d5c4dc0fd3cc04b29d3ab79c1*.

277. Philip Mobbs, "The Mineral Industry of Equatorial Guinea," Equatorial Guinea—2011 [Advance Release], *U.S. Geological Survey Minerals Handbook—2011*, p. 15.1.

278. "Equatorial Guinea," available from *www.tullowoil.com*.

279. "Production: Equatorial Guinea, Okume Complex and Ceiba Field," available from *www.hess.com*.

280. Mobbs, "The Mineral Industry of Equatorial Guinea," p. 15.1.

281. "Country Analysis Briefs—Equatorial Guinea," EIA, February 28, 2012.

282. "Equatorial Guinea," available from *www.mbendi.com/land/af/eq/p0005.htm*; and "Country Analysis Briefs—Equatorial Guinea," EIA, February 28, 2012.

283. "Country Analysis Briefs—Equatorial Guinea," EIA, February 28, 2012.

284. "Equatorial Guinea," available from *www.mbendi.com/land/af/eq/p0005.htm*;

285. "Ophir Energy Plc - Equatorial Guinea - Fortuna West-1 Update," Ophir press release, September 17, 2012.

286. "Equatorial Guinea—Drilling to Date Has Derisked 1.4 TCF," pdf file from the Ophir Energy PLC website, March 2013, available from *www.ophir-energy.com*.

287. A recent discovery made in May 2013 was at the Carla South (I-7) exploration well in Block I. See "Equa G's Carla South Exploration Well Finds Oil," *Petroleum Africa*, May 3, 2013.

288. Mobbs, "The Mineral Industry of Equatorial Guinea," p. 15.1.

289. "Noble Energy Flows Aseng, Makes New Discovery," *Petroleum Africa*, November 16, 2011; "West Africa," available from *www.nobleenergyinc.com/Operations/International/West-Africa-140.html;* and Rodney Cook, NBL Noble Energy pdf presentation, Noble Energy Analyst Conference, December 6, 2012.

290. "E&P," available from *www.glencore.com*. Another foreign partner is Sweden's PA Resources, with maximum 5.7 percent share. See "Equatorial Guinea," available from *www.paresources.se*.

291. "Equatorial Guinea," available from *www.afexglobal.com/operations/equatorial-guinea*.

292. "Train 2 Integrated Project Execution of Principles," Ministry of Mines, Industry, and Energy, press release, January 17, 2012, available from *www.equatorialoil.com*.

293. "Country Analysis Briefs—Equatorial Guinea," EIA, February 28, 2012.

294. "Signature of Production Sharing Contract for Block A-12, Offshore Bioko, Equatorial Guinea, June 18, 2012, available from *www.equatorialoil.com*.

295. "Equatorial Guinea," available from *www.marathonoil.com*.

296. "Repsol in Equatorial Guinea," available from *www.repsol.com/es en/corporacion/conocer-repsol/repsol-en-el-mundo/guinea-ecuatorial.aspx.*

297. "TOGY Talks to Tony Renton," June 20, 2012, available from *theoilandgasyear.com*; and "Equatorial Guinea," available from *www.paresources.se.*

298. "Projects—Equatorial Guinea," available from *www.panatlanticexploration.com.*

299. "CNPC in Equatorial Guinea," available from *www.chinaoil.cnpc.com.cn;* and Mario Esteban, "The Chinese Amigo: Implications for the Development of Equatorial Guinea," *The China Quarterly*, Vol. 199l, September 2009, p. 671.

300. Neil Ford, "Asian Companies Target Africa's Oil and Gas," *New African*, December 2006, p. 78; "Other Regions in Africa," available from *www.cnoocltd.com* (website indicates Equatorial Guinea assets, but does not provide detail); and Mobbs, "The Mineral Industry of Equatorial Guinea," p. 15.1.

301. "Equatorial Guinea Looks to China for Refinery," *Petroleum Africa*, July 24, 2012.

302. "Brenham Signs PSC in Equa G," *Petroleum Africa*, December 24, 2012.

303. "EG New Production Sharing Contracts—A Who's (Not) Who List," available from *transparencyafrica.com.*

304. "Murphy Optimistic on Equa Q," *Petroleum Africa*, November 1, 2012.

305. "Equa G Aims for Onshore Exploration," *Petroleum Africa*, November 9, 2011.

306. "Equatorial Guinea," available from *www.mbendi.com/land/af/eq/p0005.htm.*

307. See United Nations Security Council Resolution 2023, December 4, 2011, for details. Also see "Eritrea: UN Security Council Toughens Sanctions," BBC News, December 5, 2011.

308. "Last Stops on the East Africa Oil and Gas Frontier: Eritrea, Ethiopia," March 3, 2012, available from *oilprice.com*.

309. "Eritrea: Defba Oil Share Company Signs Oil Exploration Agreements," October 6, 2008, available from *www.energy-pedia. com*; and "Eritrean—Hidden Oil and Gas Exploration," available from *www.eritrea-chat.com*.

310. "Eritrea has Petroleum Oil and Gas," *Madote News*, October 23, 2010; and "Eritrean—Hidden Oil and Gas Exploration," available from *www.eritrea-chat.com*.

311. "Eritrean—Hidden Oil and Gas Exploration," available from *www.eritrea-chat.com*.

312. "Does Eritrea have Petroleum Oil?" *Tsfa News*, March 5, 2013.

313. "Eritrea," available from *www.mbendi.com/land/af/er/ p0005.htm*.

314. "Bid Names Tried to Make Their Mark in Play," March 26, 2010, available from *www.upstreamonline.com*.

315. "Eritrean—Hidden Oil and Gas Exploration," available from *www.eritrea-chat.com*.

316. Angelia Sanders *et al.*, "Emerging Energy Resources in East Africa," Norfolk, VA: Civil-Military Fusion Center Mediterranean Basis Team, September 2012, p. 2.

317. "Lundin Petroleum—History in Ethiopia," October 13, 2011, available from *www.lundin-petroleum.com*. Africa Oil currently has a 30 percent interest in Blocks 7 and 8 in Ogaden. See "Operations Ethiopia," available from *www.africaoilcorp.com*.

318. "Chinese Firm Picks Up Petronas' Ethiopian Assets," *Petroleum Africa*, August 17, 2011.

319. "Looting Ogaden: Ethiopia's Exploitation of Ogaden's Resources, Land and People," Resolve Ogaden Coalition, August 6, 2011, available from *www.mathaba.net*.

320. "Southwest Energy to Explore for Oil and Gas in Somali Region," August 2, 2012, available from *www.gasandoil.com*.

321. "Country Profile: Ethiopia," Cormark Securities, February 3, 2011, p. 11.

322. "Operations Ethiopia," available from *www.africaoilcorp. com*.

323. "Geology of Ethiopia—Caleb Hilala Fields," available from *https://sites.google.com/site/linkstogeologyofethiopia/Mineral/ calub-hilala-fields*.

324. "Country Profile: Ethiopia," Cormark Securities, February 3, 2011, p. 12.

325. "African Oil Farm Out in Ethiopia Closes," *Petroleum Africa*, October 29, 2012.

326. "Ethiopia," available from *www.afren.com/operations/ ethiopia/*.

327. "SouthWest Completes Ethiopia Survey," *Petroleum Africa*, February 21, 2012; and "SouthWest Plans Private Placement," *Petroleum Africa*, December 7, 2012.

328. See *sw-oil-gas.com/projects/jijiga-blocks*.

329. "Summary of Hydrocarbon Potential in the Ogaden Basin," December 2012, available from *sw-oil-gas.com/projects/ gambella-basin*.

330. See *sw-oil-gas.com/projects/jijiga-blocks*.

331. "SouthWest Receives CRP on Ethiopia Blocks," *Petroleum Africa*, April 10, 2013.

332. "Oil, Gas and Security in Ethiopia," available from *thinksecurityafrica.org/oilgas/oil-and-gas-in-Ethiopia*.

333. "Ethiopia Oil and Gas Profile," available from *abarrelfull. wikidot.com/ethiopia-oil-and-gas-profile*.

334. "Country Profile: Ethiopia," Cormark Securities, February 3, 2011, p. 10.

335. "Calvalley Signs First Production Sharing Contract in Ethiopia," October 20, 2008, available from *www.newswire.ca*.

336. See *sw-oil-gas.com/projects/gambella-basin*.

337. "Sabisa-1 Spuds in Ethiopia," *Petroleum Africa*, January 15, 2013.

338. "Africa Oil Updates Ethiopia Action," *Petroleum Africa*, April 18, 2013.

339. "Tullow Probes the 'Ramp' Prospect in Ethiopia," January 27, 2013, available from *www.africaoilgasreport.com*.

340. "Agriterra Completes Ethiopia Sale to Marathon," *Petroleum Africa*, February 1, 2013.

341. "Ethiopia," available from *www.tullowoil.com*.

342. "Africa Oil Formalizes PSA for Ethiopian Acreage," *Petroleum Africa*, February 22, 2013.

343. "Africa Oil Updates Ethiopia Action," *Petroleum Africa*, April 18, 2013.

344. "Operations Ethiopia," available from *www.africaoilcorp.com*.

345. "Calvalley Signs First Production Sharing Contract in Ethiopia," October 20, 2008, available from *www.newswire.ca*.

346. "Epsilon Would Consider Partner in Ethiopia," *Petroleum Africa*, November 3, 2011.

347. Management's Discussion and Analysis ("MD&A") for the 3 and 9 months ending September 30, 2011, p. 7, available from *www.epsilonenergyltd.com/pdf folder/Epsilon-Sept-30-2011-MDA-Final.pdf*.

348. "Ethiopia: A Potentially Golden Block on East Africa's Tertiary Rift," March 6, 2013, available from *www.oilprice.com*.

349. See *sw-oil-gas.com/projects/jimma-block*.

350. "Gabon—Country Analysis Brief," Washington, DC: U.S. Energy Information Administration, January 17, 2012.

351. "Gabon—Country Analysis Brief," EIA.

352. "Gabon: World-Class Pre-Salt Potential," *Ophir Energy. com*, p. 21.

353. See *www.mbendi.com/land/af/ga/p0005.htm*.

354. "Oil, Gas and Security in Gabon,"available from *thinksecurityafrica.org/oilgas/oil-and-gas-in-gabon/*.

355. "Gabon—Country Analysis Brief," EIA.

356. "Gabon Oil and Gas Profile," available from *abarrelfull. wikidot.com/gabon-oil-and-gas-profile*.

357. "Perenco in Gabon,"available from *www.perenco-gabon. com*.

358. "Gabon," available from *www.shell.com/global/aboutshell*.

359. "News and Media Releases," July 25, 2012, available from *www.shell.com*.

360. "Overseas oil and Gas Exploration and Development," available from *www.sinopecgroup.com*.

361. "Gabon Country Overview," available from *www.addax-petroleum.com/operations/gabon*.

362. "Gabon and Addax Locked in Dispute," *Petroleum Africa*, June 8, 2013.

363. "Upstream Africa Acreage," *Total Factbook 2011*.

364. "Marathon Oil Announces Re-entry into Gabon," press release, June 22, 2012.

365. "The Mineral Industry of Gabon," 2010 Minerals Yearbook - Gabon, U.S. Geological Survey, August 2012, p. 17.2.

366. "Vaalco Completes Avouma Well," *Petroleum Africa*, April 17, 2013.

367. "JVs/PSCs/Alliances," available from *www.oil-india.com*.

368. "Gabon" available from *www.maureletprom.fr*.

369. Saurabh Chaturvedi, "Oil India Executive: To Continue Talks for Maurel's Gabon Stake," Dow Jones, February 18, 2013.

370. "Gabon—betting on the West Coast of Africa," available from *www.petrobras.com/en/countries/gabao/gabao.htm*.

371. "VAALCO Energy Provides Update on Mutamba Iroru Block Onshore Gabon," press release, November 8, 2012, available from *vaalco.investorroom.com*.

372. "Pre-salt Discovery in Gabon," *Petroleum Africa*, January 4, 2013.

373. "Gabon: Dussafu," available from *www.harvestnr.com/operations/gabon.html*.

374. "Harvest Natural Updates Gabon Action," *Petroleum Africa*, May 6, 2013.

375. Cobalt has a 21 percent working interest in Diaba, which is operated by Total. See *www.cobaltintl.com*.

376. "Cobalt Pushes Pre-salt Program in West Africa," *Petroleum Africa*, February 28, 2013.

377. "Gabon to Launch Deepwater Licensing Round," *Petroleum Africa*, December 4, 2012.

378. "Gabonese License Awards in Offing," *AfrOil*, April 2, 2013.

379. "Pura Vida Awarded Gabon Block," *Petroleum Africa*, January 16, 2013.

380. "Pura Vida's Initial Evaluation of the Loba Discovery Well," *Petroleum Africa*, March 8, 2013.

381. See *www.africanpetroleum.com.au/our-projects/gambia*.

382. See *www.mbendi.com/land/af/gm/p0005.htm*.

383. "BOS Scores Contract for Massive Frontier Survey," *Petroleum Africa*, January 4, 2011.

384. "The Gambia," available from *buriedhill.com/Operations/The-Gambia/*.

385. "CAMAC Enters Gambia, *Petroleum Africa*, January 25, 2012.

386. CAMAC Cements Gambia Acreage," *Petroleum Africa*, May 30, 2012.

387. "Gambia," available from *www.camacenergy.com/gambia.php*.

388. "Ghana Oil Almanac," p. 13, available from *www.openoil.net*.

389. "Ghana's Upstream/Midstream Activity Map," *Africa Oil & Gas*, 2012.

390. "Jubilee Sees Record Flows," *Petroleum Africa*, December 20, 2012.

391. Ofonobong Osei-Tutu, "Kosmos Expects 25,000 BOPD (Net) in 2013," March 14, 2013, available from *africaoilgasreport.com/2013/03/in-the-news/kosmos-expects-25000bopd-net-in-2013/*.

392. "Ghana Oil Almanac," p. 13, available from *openoil.net/?wpdmact=process&did=MTYuaG90bGluaw==*.

393. *Ibid.*, p. 12.

394. "Oil, Gas and Security in Ghana," available from *thinksecurityafrica.org/oilgas/oil-and-gas-in-ghana/*.

395. "Jubilee Field," and "Deepwater Tano Block," available from *www.kosmosenergy.com*.

396. "Ghana Oil and Gas Profile," available from *abarrelfull. wikidot.com/ghana-oil-and-gas-profile;* and "Activities | Exploration & Production," available from *www.gnpcghana.com*.

397. "Activities | Exploration & Production," available from *www.gnpcghana.com*.

398. "West Cape Three Points Block," available from *www. kosmosenergy.com*.

399. "Ghana—Recent Oil Discoveries and Exploration Plans for 2011," May 23, 2011, available from *www.mergersandacquisitionreviewcom.blogspot.com*.

400. "Jubilee Field," available from *www.kosmosenergy.com*.

401. "Ghana—2010 [Advance Release]," *U.S. Geological Survey Minerals Handbook*, p. 19.4. Note: A different source gave the shares in Jubilee as Tullow, 36.5 percent; Anadarko and Kosmos, 23.49 percent each; GNPC, 13.75 percent; and Sabre, 2.81 percent. See "Oil & Gas—Ghana and Uganda—the Quick and the Slow," November 18, 2012, available from *www.equities.com*.

402. "Major Projects" available from *www.tullowoil.com*.

403. "Tullow Looks to Sell a Piece of TEN," *Petroleum Africa*, February 15, 2013.

404. "TEN PoD Gets Nod from Ghana," *Petroleum Africa*, May 31, 2013.

405. "New Discovery May Lead to Fourth Field Development Offshore Ghana," *Africa Oil & Gas Journal*, August 8, 2012; and "Jubilee Sees Record Flows," *Petroleum Africa*, December 20, 2012.

406. Note: Vitol's website indicates the firm has a 44.44 percent non-operator interest, with ENI holding a 55.56 percent interest and GNPC a carried 15 percent interest. See "Ghana," available from *www.vitol.com*.

407. "ENI Enlarges the Ghanaian Oil Tank," *Africa Oil & Gas Report*, October 9, 2012.

408. "Ghana's Upstream/Midstream Activity Map," *Africa Oil & Gas*, 2012.

409. "ENI Discovers Oil in Tano Basin Offshore Ghana," September 20, 2012, available from *www.epmag.com*.

410. Alan Petzet, "Ghana discovery list grows with Vanco-Lukoil Find," March 15, 2010, available from *www.ogj.com*.

411. "Ghana—Recent Oil Discoveries and Exploration Plans for 2011," May 23, 2011, available from *www.mergersand acquisitionreviewcom.blogspot.com*.

412. "Ghana," available from *www.statoil.com/Annual Report2011/en/Pages/frontpage.aspx;* and "Hess Gets Partner in Ghana," *Petroleum Africa*, April 27, 2012.

413. "Deepwater Tano/Cape Three Points," available from *hess.com/operations/default.aspx*.

414. "Ghana's Offshore Pecan-1 a Hit for Hess," *Petroleum Africa*, December 13, 2012.

415. "Hess Cracks Nut Discovery Offshore Ghana," *Petroleum Africa*, March 1, 2013.

416. Alan Petzet, "Vanco, Lukoil find oil off eastern Ghana," February 26, 2010, available from *www.ogj.com*.

417. "Lukoil Could Resume Ghana Drilling in 4Q," *Petroleum Africa*, March 29, 2013.

418. "Lukoil Eyes New Ghanaian Drilling," *AfrOil*, April 2, 2013.

419. "Aker wins deepwater acreage in Ghana," available from *www.akerasa.com/News-Media/Stock-exchange-releases/Agreement/Aker-wins-deepwater-acreage-in-Ghana/(year)/2013.*

420. "Cape Three Points Deep Water Block," November 6, 2008, available from *www.panatlanticexploration.com.*

421. "Offshore Saltpond Basin—Gasop," available from *www.gasopoil.com.*

422. "Hydrocarbon exploration in Ghana," available from *www.tullowoil.com.*

423. See *en.wikipedia.org/wiki/Saltpond Oil Field.*

424. "Keta Block," available from *www.afren.com/operations/ghana/keta block.*

425. "Ghana's Upstream/Midstream Activity Map," *Africa Oil & Gas*, 2012.

426. "Keta Block," available from *www.afren.com/operations/ghana/keta block.*

427. "Ghana's Upstream/Midstream Activity Map," *Africa Oil & Gas*, 2012.

428. "Ghana - Offshore Accra Block," available from *www.rialtoenergy.com/index.php?option=com content&view=article&id=27&Itemid=27.*

429. "Stena signed up for more Ghanaian Drilling," *AfrOil*, April 9, 2013.

430. "Simba Energy in Ghana," available from *www.simbaenergy.ca/projects/ghana.aspx.*

431. "Oil & Gas—Ghana and Uganda—the quick and the slow," available from *www.equities.com.*

432. "Ghana's Upstream/Midstream Activity Map," *Africa Oil & Gas*, 2012.

433. "Jubilee Gas Tagged for Aboadze Power," *Petroleum Africa*, April 2, 2013.

434. "Ghana-Ivory Coast Border Dispute Doesn't Distract Us—Kosmos Energy," September 27, 2012, available from *www.ghanaoilwatch.org*.

435. "Ivory Coast-Ghanaian Border Dispute," *en.wikipedia.org*.

436. "Oil, Gas and Security in Guinea," available from *thinksecurityafrica.org/oilgas/oil-and-gas-in-Guinea*.

437. "Guinea," available from *www.dana-petroleum.com/Our-businesses/Africa/Overview*.

438. "West Africa Operations," available from *hyperdynamics.com/guinea project.htm*.

439. *Ibid.*

440. Guinea—Country Commercial Guide 2012," U.S. Foreign Commercial Service.

441. See *www.chinasonangol.com*.

442. "Hyperdynamics Announces Updated Prospective Oil Resources Evaluation," February 14, 2013, available from *investors.hyperdynamics.com/releasedetailcfm?ReleaseID=740406*. and "Tullow Agrees to Farm-in Guinea Concession," November 20, 2012, available from *www.tullowoil.com*.

443. "HyperD's Guinea Farm-out to Tullow a Done Deal," *Petroleum Africa*, January 2, 2013.

444. "Guinea: Hyperdynamics Encounters Oil Shows in Sabu-1 Well Offshore Guinea-Conakry," February 15, 2012, available from *www.energy-pedia.com*.

445. "Simba Lands Onshore Guinea," *Petroleum Africa*," July 29, 2011.

446. "Simba Gets Green Light for Onshore Guinea PSCs," *Petroleum Africa*, May 25, 2012.

447. "Simba Energy Identifies Three Significant Seeps in Guinea," February 27, 2013, available from *www.simbaenergy.ca*.

448. "Guinea, Bove Basin," available from *www.simbaenergy.ca/projects/guinea.aspx*.

449. "Guinea Hammers Out Petroleum Code-Auction in the Works," *Petroleum Africa*, November 2, 2012.

450. "Oil, Gas and Security in Guinea-Bissau," available from *thinksecurityafrica.org/oilgas/oil-and-gas-in-Guinea-Bissau*.

451. "Guinea-Bissau," available from *www.mbendi.com/land/af/gb/p0005.htm*.

452. David Brown, *The Challenge of Drug Trafficking to Democratic Governance and Human Security in West Africa*, Carlisle, PA: Strategic Studies Institute, U.S. Army War College, May 2013.

453. "Svenska Finalizes Data Acquisition Offshore Guinea Bissau," Svenska Petroleum press release, December 10, 2010.

454. *Ibid.*

455. "Guinea-Bissau," available from *www.far.com.au/guinea-bissau/*.

456. See *www.spherepetroleum.com/Guinea-Bissau/guinea-bissau.html*.

457. "Noble Joins Senegal-Guinea Bissau Exploration," *Oil & Gas Journal*, June 8, 2011.

458. "Completion of Kora-1 Exploration Well," Ophir press release, July 27, 2011.

459. ""AGC," available from *www.far.com.au*.

460. "Country Profile: Kenya," Toronto, Canada: Cormark Securities, February 3, 2011, p. 15.

461. Eduard Gismatullin, "Kenya's Deepwater Debut Heralds East Africa's First Oil: Energy," Bloomberg, July 11, 2012.

462. "Kenya Block 2A," available from *www.simbaenergy.ca.*

463. "Country Profile: Kenya," Cormark Securities, February 3, 2011, p. 15.

464. "Kenya to Gazette up to 9 New Oil and Gas Exploration Blocks," *Reuters*, February 12, 2013.

465. "Cradle of Mankind and the Next Big Oil Patch," *Bloomberg BusinessWeek*, March 18-24, 2013, p. 10.

466. "Kenya," March 16, 2013, available from *www. camacenergy.com/kenya.php;* Gismatullin, "Kenya's Deepwater Debut Heralds East Africa's First Oil: Energy."

467. Angelia Sanders *et al.*, "Emerging Energy Resources in East Africa," Norfolk, VA: Civil-Military Fusion Center Mediterranean Basis Team, September 2012, p. 3.

468. "Kenya to Gazette up to 9 New Oil and Gas Exploration Blocks," *Reuters*, February 12, 2013.

469. Sarah McGregor, "Kenya Spends $25 Billion on Bond-Backed Port for Oil: Freight," September 4, 2012, available from *Bloomberg.com.*

470. "Twiga South-1 Hits Oil," *Petroleum Africa*, November 1, 2012; and "Kenya's Twiga South-1 Flow tests Complete," *Petroleum Africa*, February 22, 2013.

471. "Kenya: Oil Production 6-7 Years Away," *Oilprice.com*, May 10, 2013.

472. "Kenya," available from *www.tullowoil.com.*

473. "Block 12B, Kenya," available from *www.swala-energy. com.*

474. "Paipai-1 Well Suspended," *Petroleum Africa*, March 4, 2013.

475. "Africa Oil Completed Deal with Marathon Oil on Kenya Blocks," available from *www.equities.com*.

476. "Kenya," available from *www.africaoilcorp.com*.

477. Sanders *et al.*, "Emerging Energy Resources in East Africa," p. 2.

478. "Seismic Bulks Up Afren's East African Resource Portfolio," *Petroleum Africa*, August 23, 2013.

479. "Overview," available from *taipanresources.com/ operations-overview.html*.

480. "Kenya Taipan Resources Completes 2D Seismic Survey in Block 2B onshore Kenya—Commences FTG Survey," March 21, 2013, available from *www.energy-pedia.com/news/kenya/ new-153949*.

481. "Projects—East Africa Opportunity," available from *www.imaraenergy.com/projects/index.php*.

482. "Overview," available from *taipanresources.com/ operations-overview.html*.

483. "ERHC Signs PSC for Kenyan Acreage," *Petroleum Africa*, July 10, 2012.

484. "Republic of Kenya," available from *www.erhc.com*.

485. "ERHC Signs LoI for Farm Out in Kenya," *Petroleum Africa*, May 9, 2013.

486. "Kenya, Block 2A," available from *www.simbaenergy.ca/ projects/kenya.aspx*.

487. "Africa Oil Completes Deal with Marathon Oil on Kenya Blocks," available from *refinerynews.com/africa-oil-completes-deal-with-marathon-oil-on-kenya-blocks/*.

488. "Simba to Farm Out Kenyan Stake," *Petroleum Africa*, May 14, 2013.

489. "Oil Industry in Kenya," available from *www.vanoil.ca/s/Kenya.asp*.

490. "Kenya Oil and Gas Profile," available from *abarrelfull.wikidot.com/kenya-oil-and-gas-profile*.

491. "Kenya Assets," available from *www.bowleven.com*.

492. "Kenya" [Advance Release], *U.S. Geological survey Minerals Yearbook 2010*, p. 22.2. Note: One source indicated that CNOOC left Kenya after this unsuccessful well, but the firm's website indicates, without providing any detail, that it still has operations in Kenya. See *www.cnoocltd.com*; "Country Profile: Kenya," Cormark Securities, February 3, 2011, p. 17.

493. "Kenya," available from *investor.apachecorp.com/releasedetail.cfm?ReleaseID=706491*.

494. Sanders *et al.*, "Emerging Energy Resources in East Africa," p. 2.

495. "Kenya—Lamu Basin," available from *www.pancon.com.au/projects/kenya/lamu-basin-l8-l9/*.

496. "Kenya L5, L7, L11A, L11B & L12 Project," available from *www.pttep.com*.

497. "Kenya," March 16, 2013, available from *www.camacenergy.com/kenya.php*.

498. "CAMAC Commences Kenyan Surveys," *Petroleum Africa*, April 11, 2013.

499. "FAR Completes 3D Data Acquisition Offshore Kenya," July 12, 2012, available from *www.subseaworldnews.com*; "Kenya—Lamu Basin," available from *www.pancon.com/au*; and "FAR Sees Block L6 Expanded," *Petroleum Africa*, March 28, 2013.

500. "FAR expands L-6 block off Kenya," *AfrOil*, April 2, 2013.

501. "FAR Keen on Kenya/Senegal Drilling Prospects," *Petroleum Africa*, February 28, 2013. Also see *www.far.com.au/kenya/*.

502. "Kenya: Ophir Operated Drilling Programme," available from *www.ophir-energy.com*.

503. "Kenya," available from *www.bg-group.com*.

504. "Kenya," available from *www.premier-oil.com*.

505. "Block L17/L18," available from *www.afren.com/operations/kenya/block l17l18/*.

506. "Afren Tangos in Tanzania's Tango Block," *Petroleum Africa*, March 25, 2011.

507. "Seismic Bulks Up Afren's East African Resource Portfolio," *Petroleum Africa*, August 23, 2013.

508. "Mineral Fuels"; [Advance Release] *U.S. Geological Survey – 2011*; "Second Discovery Points Liberia As Emerging Western Star," *Africa Oil & Gas Report*, March 5, 2013.

509. "Repsol in Liberia," available from *www.repsol.com/es en/corporacion/conocer-repsol/repsol-en-el-mundo/liberia.aspx*.

510. "Liberia," available from *www.chevron.com*; and Christopher Helman, "Chevron Adds Big Acreage Offshore Liberia," *Forbes*, September 8, 2010.

511. "Liberia on Course to Prove Oil Riches in 2011," June 2011, available from *mergersandacquisitionreviewcom.blogspot.com*.

512. "Updates West Africa Activity," news release, November 9, 2011, available from *www.anadarko.com*.

513. "Liberia," available from *www.tullowoil.com/index.asp?pageid=267*.

514. "African Petroleum Launches Liberia Drilling, *Petroleum Africa*, January 9, 2013.

515. "Liberia's Bee Eater-1 a Hit for AP," *Petroleum Africa*, February 20, 2013.

516. "Oil, Gas and Security in Liberia," available from *thinksecurityafrica.org/oilgas/oil-and-gas-in-Liberia*.

517. "Liberia," available from *www AfricanPetroleum.com.au/our-projects/liberia*.

518. "ExxonMobil Dives in Offshore Liberia," *Petroleum Africa*, March 11, 2013; and "Liberia to Ratify PSC for LB-13," *Petroleum Africa*, March 28, 2013.

519. "Liberia ratifies Canadian Overseas' PSC," *AfrOil*, April 2, 2013.

520. "TGS Launches Liberian Campaign," *Petroleum Africa*, accessed January 22, 2013.

521. "Gas and Security in Liberia," available from *thinksecurityafrica.org/oilgas/oil-and-gas-in-Liberia*.

522. "Simba Energy in Liberia," available from *www.simbaenergy.ca/projects/liberia.aspx*.

523. "Libya Oil and Gas Profile," available from *www.abarrelfull.wikidot.com*.

524. "Oil, Gas and Security in Libya," available from *thinksecurityafrica.org/oilgas/oil-and-gas-in-Libya*.

525. "Country Analysis Briefs—Libya," EIA, June 2012.

526. Taib, "The Mineral Industry of Libya," p. 25-1.

527. Christopher Coats, "Libya Protection Falling Short," *AfrOil*, April 2, 2013.

528. "Country Analysis Briefs—Libya," EIA, June 2012.

529. Taib, "The Mineral Industry of Libya," p. 25-1.

530. "Country Analysis Briefs – Libya," EIA, June 2012.

531. Taib, "The Mineral Industry of Libya," p. 25-1.

532. "Country Analysis Briefs – Libya," EIA, June 2012.

533. Taib, "The Mineral Industry of Libya," p. 25-1.

534. "Exploration: Libya Waha," available from *www.hess.com/operations*; and "Libya," available from *www.marathonoil.com/Global Operations/Libya*.

535. "Libya - Sirte Basin Waha Concession," available from *www.conocophillips.com/EN/about/worldwide ops/other/Pages/Libya*.

536. "Exploration: Libya Area 54," available from *www.hess.com/operations*.

537. "Repsol in Libya," available from *www.repsol.com/es en/corporacion/conocer-repsol/repsol-en-el-mundo/libia.aspx*.

538. Esam Mohamed, "Militiamen Attack Ga Complex in Western Libya," Associated Press, May 20, 2013.

539. "Wintershall in Libya," available from *www.wintershall.com/en/worldwide/libya.html*.

540. "Libya," available from *www.gazprom.com/about/production/projects/deposits/libya*.

541. "Libya," available from *www.suncor.com*.

542. "Libya," available from *www.statoil.com/Annual Report2011/en/Pages/frontpage.aspx*.

543. "Upstream Africa Acreage – Libya," *Total Factbook 2011*; and "Libya," "Exploration and Production in Africa – Total Group," available from *www.total.com*.

544. "Wintershall in Libya," available from *www.wintershall. com/en/worldwide/libya.html.*

545. "Libya," available from *www.omv.com/portal/01/com/omv/ OMVgroup/Business Segments/OMV Exploration and Production/ Worldwide Activities/Libya.*

546. "Medco Energi Establishes JOC for Libya Ops," *Petroleum Africa*, March 26, 2013.

547. "Libya Oil Development," available from *www.med coenergi.com/page.asp?id=210028.*

548. "JVs/PSCs/Alliances," available from *www.oil-india.com.* Note: Oil India's website indicates this NOC is in the process of relinquishing area 86 and Clock 102(4).

549. "CNPC in Libya," available from *www.cnpc.com.cn/en/ cnpcworldwide/libya.*

550. See *www.gwdc.com.cn.*

551. "Oil giant Exxon returns to Libya," December 6, 2005, available from *www.bbc.co.uk.*

552. "Shell Says Still Interested in Libya Oil Exploration," *Reuters*, November 7, 2012.

553. See *www.upstreamonline.com*, October 4, 2010.

554. Christopher Coats, "Libya Protection Falling Short," *AfrOil*, April 2, 2013.

555. "Libyan protesters end blockage at Waha Oil field," *AfrOil*, April 2, 2013.

556. Debbie Stevenson, "Investors Make a Move on Madagascar's Oil Reserves," November 20, 2012.

557. "Country Profile: Madagascar," Cormark Securities, February 3, 2011, p. 19.

558 . "Oil, Gas and Security in Madagascar," available from *thinksecurityafrica.org/oilgas/oil-and-gas-in-madagascar/*.

559. "Madagascar Puts Kibosh on Oil Tender," *Petroleum Africa*, March 7, 2011.

560. Stevenson, "Investors Make a Move on Madagascar's Oil Reserves."

561. "Uncertainty Reigns for Madagascar Oil, *Petroleum Africa*, February 10, 2011. See also "Country Profile: Madagascar," Cormark Securities, February 3, 2011, p. 18.

562. "Madagascar Puts Kibosh on Oil Tender," *Petroleum Africa*, March 7, 2011.

563. "Force Majeure for Madagascar Oil," *Petroleum Africa*, March 22, 2011.

564. "Madagascar Settles Dispute Over Tsimiroro Block," *Petroleum Africa*, June 27, 2011.

565. "Madagascar Oil Sees Resolution with OMNIS," *Petroleum Africa*, April 6, 2012.

566. "Madagascar Oil See Tsimiroro Resources Rise," *Petroleum Africa*, September 13, 2011; and "Focused on the Development of Heavy Oil & and Conventional Oil & Gas," available from *www.madagascaroil.com*.

567. Madagascar Oil Limited, Securities Announcement of Half Year Results, September 28, 2012.

568. EIA Short-Term Energy Outlook (STEO), as provided by USAFRICOM source, April 25, 2013.

569. Stevenson, "Investors Make a Move on Madagascar's Oil Reserves."

570. "Madagascar," available from *www.total.com*.

571. "Total's Stay Extended on Madagascar Block," *Petroleum Africa*, July 29, 2011.

572. "Madagascar," January 17, 2013, available from *www. afren.com/operations/madagascar.*

573. "Afren Grabs Huge Madagascar Stake," *Petroleum Africa,* July 26, 2011.

574. "Madagascar," January 17, 2013, available from *www. afren.com/operations/madagascar.*

575. "Caravel launches Seismic in Madagascar," *Petroleum Africa,* December 17, 2013.

576. "Petrole et gaz a Madagascar—Behaza Oil Project . . ." ("Oil and Natural Gas in Madagascar—Behaza Oil Project . . ."), December 19, 2012, available from *www.tim-madagascar.org.*

577. "Core Assets—Madagascar," available from *www.avana-petroleum.com/islands.*

578. "Tullow Looking for Partner in Madagascar," November 4, 2011.

579. "Madagascar," available from *www.tullowoil.com.*

580. Overseas assets, available from *www.essarenergy.com.*

581. "Madagascar—Exxon Keeps Eye on Offshore," January 18, 2012, available from *www.africaintelligence.com.*

582. "Madagascar," Vietnam National Oil and Gas Group, available from *english.pvn.vn;* "Madagascar," available from *www. bg-group.com.*

583. "Madagascar: Ampasindava," available from *www. sterlingenergyuk.com.*

584. "Country Profile: Madagascar," Cormark Securities, February 3, 2011, p. 19.

585. "Madagascar," available from *nikoresources.com/ operations/madagascar.html.*

586. "Completion of 2-D Seismic Data Acquisition in Juan de Nova and Belo Profond Assets," press release, available from *www.sapetro.com*.

587. "SAPETRO Purchases 90% Interest in Belo Profond Off-shore Madagascar," press release, October 27, 2011, available from *www.sapetro.com*.

588. "Completion of 2-D Seismic Data Acquisition in Juan de Nova and Belo Profond Assets."

589. "Juan de Nova," available from *www.globalpetroleum.com. au/operations/juan-de-nova*.

590. "Juan de Nova Est Mozambique Channel," available from *www.wessexexploration.com/JuanDeNova.html*.

591. Paul Burkhardt, "SacOil Wins Malawi Block as East African Finds Lure Drillers," Bloomberg, December 13, 2012.

592. "Award of the Onshore Petroleum Prospecting License Block 1 in Malawi," December 13, 2012, available from *www. sacoilholdings.com*.

593. "Surestream Petroleum Awarded Exploration Licenses in Malawi," September 22, 2011, available from *www.surestream-petroleum.com*.

594. Robyn Curnow *et al.*, "Troubled Water: Oil Search Fuels Tension Over Lake Malawi," CNN, November 14, 2012.

595. Tanzania Calls on Malawi to Quit Exploring," *Petroleum Africa*, July 31, 2012; "Tanzania Wants Mediator for Lake Malawi Dispute," *Petroleum Africa*, October 8, 2012.

596. "Malawi/Tanzania Dispute Widens with New Award," *Petroleum Africa*, March 1, 2013.

597. "Malawi, Tanzania Head for Court," *AfrOil*, April 9, 2013.

598. Abdul Wakil Saiboko, "Tanzania: Lake Nyasa Border Dispute Goes to African Forum," *Tanzania Daily News*, Decem-

ber 29, 2012; and "Malawi/Tanzania Dispute Widens with New Award," *Petroleum Africa*, March 1, 2013.

599. "Heritage Scores Another Tanzania PSA," *Petroleum Africa*, January 26, 2012.

600. "NuEnergy Granted Exclusive Prospecting Licence for CBM and Shale Gas Exploration in Malawi," NuEnergy Company announcement, July 9, 2012.

601. "NuEnergy Launches Malawi Airborne Survey," *Petroleum Africa*, May 17, 2013.

602. "Oil, Gas and Security in Mali," available from *www.thinksecurityafrica.org/oilgas/oil-and-gas-in-Mali*.

603. "Mali," available from *afexglobal.com/operations/mali*.

604. "Mali—A Developing Oil and Gas Industry," Mali Republic Minister of Mines, Energy and Water, presentation to Corporate Council on Africa conference, November 29, 2006.

605. "ENI Pulls Out of Mali on Poor Prospecting Outlook," *Reuters*, January 15, 2013.

606. "Angola Enters Mali," *Petroleum Africa*, November 8, 2011.

607. "Simba Receives PSA Approval from Mali," *Petroleum Africa*, October 2011.

608. Homepage, available from *www.spherepetroleum.com*.

609. Map, "Simba Energy in Mali," available from *www.simbaenergy.ca/projects/mali.aspx*.

610. Pav Jordan, "Iamgold to Cut Back Mali Exploration Activity," *Globe and Mail*, January 23, 2013.

611. "A propos de Petroma" ("About Petroma"), available from *www.petroma-mali.com/fr/*.

612. "Mali," available from *afexglobal.com/operations/mali*.

613. "Mauritania, Benin, and Togo: Overview of Oil and Gas Discoveries and Exploration in 2011," available from *mergersandacquisitionreviewcom.blogspot.com/2011/06/mauritania-benin-and-togo-overview-of.html*.

614. "Oil, Gas and Security in Mauritania." available from *www.thinksecurityafrica.org*.

615. "Mauritania, Benin, and Togo: Overview of Oil and Gas Discoveries and Exploration in 2011," available from *mergersandacquisitionreviewcom.blogspot.com/2011/06/mauritania-benin-and-togo-overview-of.html*.

616. "Mauritania," available from *www.tullowoil.com*.

617. "CNPC in Mauritania," available from *www.cnpc.com.cn/en/cnpcworldwide/mauritania*.

618. ""Update on DNPCIM-operated Heron-1 in Mauritania," *Petroleum Africa*, January 17, 2007.

619. "Successful Cormoran-1 Exploration Well Offshore Mauritania," available from *www.tullowoil.com*.

620. "Mauritania," available from *www.tullowoil.com*.

621. "12 month Exploration & Appraisal Programme," available from *www.tullowoil.com*.

622. "Kosmos Enters Mauritania," *Petroleum Africa*, April 10, 2012.

623. "Overview — Mauritania," available from *www.chariotoilandgas.com/operations/mauritania*.

624. "Total Awarded Two Exploration Licenses in Mauritania, Offshore and Onshore," news release, January 6, 2012, available from *www.total.com*.

625. "CNPC in Mauritania," available from *www.cnpc.com.cn/en/cnpcworldwide/mauritania*.

626. "Mauritania," available from *www.rwe.com*.

627. "Repsol in Mauritania," available from *www.repsol.com*.

628. "Total Awarded two Exploration Licenses in Mauritania, Offshore and Onshore," news release, January 6, 2012, available from *www.total.com*.

629. "India to Explore for Oil, Gas Off Mauritius," July 3, 2003, available from *www.siliconindia.com*.

630."Mauritius: ONGC Considering Oil and Gas Exploration Rights," July 8, 2007, available from *www.energy-pedia.com/news/mauritius/ongc-considering-oil-and-gas-exploration-rights-*.

631. "Historique de l'Exploitation" ("History of Exploration"), Rabat, Morocco : Office National des Hydrocarbures et des Mines, available from *www.onhym.com/Hydrocarbons/Explorationhistory/tabid/290/Default.aspx?Cat=3*.

632. "Morocco," available from *www.mbendi.com/land/af/mo/p0005.htm*.

633. "Circle Oil Lays Out North Africa Program," *Petroleum Africa*, February 20, 2013.

634. "Activites d'Exploration" ("Exploration Activities"), Rabat, Morocco: Office National des Hydrocarbures et des Mines, available from *www.onhym.com/HYDROCARBURES/Activit%C3%A9sdexploration/Propre%C3%A0lONHYM/tabid/356/language/en-US/Default.aspx?Cat=28*.

635. "Morocco," available from *www.sanleonenergy.com/operations-and-assets.aspx/#morocco*; and "Award of Additional Blocks," San Leon Energy press release, August 29, 2012.

636. "Oil and Gas," available from *en.wikipedia.org/wiki/energy in Morocco*.

637. "Operations in Morocco," available from *www.chariotoilandgas.com*.

638. "Chevron Lands Deep Offshore Morocco," *Petroleum Africa*, January 23, 2013.

639. "Morocco Significant New Play Opportunities," available from *www.kosmosenergy.com/operations-morocco.php*.

640. "Kosmos' 3Q Results Are In," *Petroleum Africa*, November 6, 2012.

641. "Farmout Secures Funding for Multi-Well Drilling Program Offshore Morocco," Puravida Energy press release, January 3, 2013.

642. Based on various company websites, including *tangierspetroleum.com, repsol.com, gulfsands.com, longreachoilandgas. com, cairnenergy.com, chariotoilandgas.com, galpenergia.com, sericaenergy.com*. See also "Pura Vida Enters Morocco," *Petroleum Africa*, November 21, 2011; and *otp.investis.com*.

643. "Repsol in Morocco," available from *www.repsol.com/ es en/corporacion/conocer-repsol/repsol-en-el-mundo/marruecos.aspx*.

644. "Presence of Oil and Gas in Morocco," available from *thinksecurityafrica.org/oilgas/oil-and-gas-in-morocco/*.

645. "Mozambique: Oil & Gas Exploration in 2011-2012," available from *mergersandacquisitionreviewcom.blogspot.com/2011/ 07/mozambique-oil-gas-exploration-in-2011.html*.

646. "Mozambique Expects Megabillions for LNG," *Petroleum Africa*, February 9, 2012.

647. See *www.anadarko.com/Operations/Pages/LNGmozambique. aspx*.

648. "Mozambique," available from *www.anadarko.com/ Operations/Pages/LNGmozambique.aspx*.

649. "Anadarko Reaches Heads of Agreement with ENI, Award Front-end Engineering and Design Contracts," Anadarko News Release, December 21, 2012.

650. "Mozambique Fact Sheet 2013," Anadarko Petroleum.

651. "Anadarko Considers JV for Mozambique Monetization," *Petroleum Africa*, November 29, 2012.

652. "Indian Firms May Ante up for Mozambique Offshore Area 1 Stake," *Petroleum Africa*, March 15, 2013.

653. "Anadarko Eyes Japan for Mozambique LNG Sales," *Petroleum Africa*, March 14, 2013.

654. "Mozambique: Another Success on Offshore Area 1," *Petroleum Africa*, January 17, 2012, available from *www.pttep.com*.

655. "Oil and Gas—Mergers and Acquisition Review: Mozambique Oil & Gas Exploration in 2011," available from *mergersandacquisitionreviewcom.blogspot.com/2011/07/mozambique-oil-gas-exploration-in-2011.html*.

656. "ENI Adds Trillions to Mozambique Bounty," *Petroleum Africa*, December 6, 2012; "ENI's Coral-3 Adds to Area 4 Bounty," *Petroleum Africa*, February 26, 2013.

657. Sa'd Bashir, "East Africa on Sale," *Africa Oil & Gas Report*, May 21, 2013.

658. Justin Scheck *et al.*, "ENI, CNPC Link Up on Mozambique Gas," *Wall Street Journal Europe*, March 15-17, 2013.

659. "Statoil Farms Down in Mozambique License," August 14, 2012, available from *www.statoil.com/en/NewsAndMedia/News/2013/Pages/02Apr Mozambique.aspx*. "Mozambique," available from *www.tullowoil.com*.

660. "Statoil Sells Down Mozambique Stake to Inpex," *Petroleum Africa*, April 3, 2013.

661. "Mozambique: Petronas, Total to Explore Offshore Mozambique," *Oil & Gas Journal*, September 24, 2012.

662. "Mozambique," available from *www.mbendi.com/land/af/mz/p0005.htm.*

663. "Mozambique Third License Round, Technical Overview," Maputo, Mozambique: Instituto Nacional de Petroleo, June 2008.

664. "Mozambique—2010 (Advance Release)," *U.S. Geological Survey—Minerals Handbook 2010*, p. 31.3.

665. "South Africa's Sasol Analyzes Condensate Deposits in Inhambane, Mozambique," July 9, 2012, available from *www.macauhub.com.mo/en/2012/07/09/south-africa%e2%80%99s-sasol-analyses-condensate-deposits-in-inhambane-mozambique/.*

666. "Rovuma Onshore Concession," available from *www.wentworthresources.com*; "Mozambique Rovuma Onshore Project," available from *www.pttep.com*.

667. "Doing Business in Mozambique: 2011 Country Commercial Guide for U.S. Companies," Brussels, Belgium: U.S. Foreign Commercial Service.

668. "Oil, Gas and Security in Mozambique," available from *thinksecurityafrica.org/oilgas/oil-and-gas-in-Mozambique.*

669. Anne Fruhauf, "Mozambique's Gas Sector: Prospects and Perils," Oxford Energy Forum, Oxford, UK, November 2012, p. 12.

670. *Ibid.*

671. "Mozambique Enacts New Tax," *Petroleum Africa*, December 20, 2012.

672. "Mozambique to Raise Its Stake Participation," *Petroleum Africa*, April 30, 2012.

673. "Mozambique's Master Plan Ready by Year-end," *Petroleum Africa*, May 9, 2013.

674. "Petrole: Guebuz ecarte la possibilite d'un conflit avec la Tanzanie" ("Petroleum: Guebuz Dismisses the Possibility of a Conflict with Tanzania"), *Xinhua*, December 13, 2011.

675. "Chariot/BP Namibia Deal Done," *Petroleum Africa*, February 3, 2012.

676. "Namibia," available from *www.galpenergia.com*.

677. "Namibian Block 0010 Partners Win Farm-Out Approval," *Petroleum Africa*, December 24, 2012.

678. Brian Swint, "Chariot Oil & Gas Plans More Namibia Wells After Two Dry Holes," Bloomberg, September 21, 2012.

679. "Oil, Gas and Security in Namibia," available from *thinksecurityafrica.org/oilgas/oil-and-gas-production-in-namibia/*.

680. "Overview Namibia," "Exploration Detail—Namibia," available from *www.chariotoilandgas.com*.

681. "Namibia," available from *www.mbendi.com/land/af/na/ p0005.htm*.

682. "Regions Petroliferes en Afrique" ("Petroleum Regions in Africa"), available from *fr.wikipedia.org*.

683. "Namibia," available from *www.tullowoil.com*.

684. "Namibia—2010" (Advance Release), *U.S. Geological Survey Minerals Yearbook*, p. 32.3. One news report indicated that Gazprom quit the Kudu project, leaving Namcor to seek a replacement investor. See "Upstream Update: Upsurge in Namibia and South Africa," *African Energy*, May 2, 2013.

685. "HRT Heralds Return of the Namibian Rush," *Africa Oil & Gas Report*, March 5, 2013; and "HRT spuds Offshore Namibia," *Petroleum Africa*, March 27, 2013.

686. "HRT Prepares for Next Namibian Spud," *Petroleum Africa*, May 24, 2013.

687. "Global Opens Data Room for Namibia License," *Petroleum Africa*, January 15, 2013.

688. "Walvis Bay, Offshore Northern Namibia," available from *www.pancon.com.au/projects/namibia/walvis-basin*.

689. "Company Profile," available from *www.ecooilandgas.com*.

690. "18 Block," "19'S Block," available from *www.westbridgeweb.com/s/18 block.asp; and www.westbridgeweb.com/s/19 block.asp*.

691. "Two Exploration Licenses Awarded by Government of Namibia," No. 07-12, Paris, France: Maurel et Prom, March 15, 2012.

692. "New Farm Out Offshore Namibia," *Petroleum Africa*, May 2, 2013. Oranto farmed Bermuda-based Acorn Geophysical in May 2013 into its stakes in 2011B/2111A offshore permit in Walvis Bay.

693. "Signet Petroleum Ltd.," available from *www.poloresources.com/Investments Signet.htm*.

694. "Namibia Approves Serica Energy's Farm Out," *Petroleum Africa*, June 27, 2012, available from *www.serica-energy.com*.

695. "Tower's Namibia Deal Approved," *Petroleum Africa*, February 1, 2013.

696. "EnerGulf Enters Second Renewal in Namibia," *Petroleum Africa*, July 31, 2012, available from *www.petroleumafrica.com/energulf-enters-second-renewal-in-namibia/*; Block 1711, available from *www.energulf.com/assets/namibia/*.

697. "Pancontinental's Namibian Block has Big Oil Potential," *Petroleum Africa*, May 28, 2013.

698. "19's Block," available from *www.westbridgeweb.com/s/19 block.asp*.

699. "Global Reports on Namibia and Juan de Nova," *Petroleum Africa*, April 29, 2012.

700. "Namibia: The Next Hot Spot for Oil Exploration in West Africa?" *Seeking Alpha*, October 26, 2011.

701. "Namibia Piles Up the Reserves," *Petroleum Africa*, July 7, 2011.

702. Josh Lewis, "Petrobras Tastes Dust of Namibia," September 10, 2012, available from *upstreamonline.com*.

703. "Nimrod a No-Go for Chariot and Partners," *Petroleum Africa*, September 11, 2012.

704. "Namibia: The Next Hot Spot for Oil Exploration in West Africa?"

705. *Ibid.*; "Namibia: Tapir South Disappoints—Nimrod Next," *Petroleum Africa*, May 15, 2012.

706. "Republic of Namibia," available from *www.duma.com/projects/namibia*.

707. "Namibia: Owambo Basin Structural Geology Studied," *Oil & Gas Journal*, available from *www.ogj.com/articles/2012/02/namibia-owambo-basin-structural-geology-studied.html*.

708. "HydroCarb Energy Updates Source rock Evaluation for its Africa Concession," November 27 2012, available from *www.hydrocarb.com*.

709. "Duma's Namibia P10 Resources Rise," *Petroleum Africa*, December 5, 2012.

710. "Projects: Namibia," available from *www.friplc.com/index.php/operations/namibia*.

711. "Upstream update: Zambia, Malawi, and Botswana," *African Energy*, May 16, 2013.

712. "Company Profile," available from *www.ecooilandgas.com*.

713. "Nigerien Hydrocarbon Reserves," available from *wiki.openoil.net*.

714. "Oil, Gas and Security in Niger," available from *thinksecurityafrica.org/oilgas/oil-and-gas-in-the-republic-of-niger/*.

715. "Nigerian Hydrocarbon Reserves," available from *wiki.openoil.net*.

716. "Sonatrach Operations in Niger," available from *wiki.openoil.net*.

717. "CNCP in Niger," available from *www.cnpc.com.cn/en/cnpcworlwide/niger/*.

718. "CNPC and CPC Corp. To Jointly Explore Niger," *Petroleum Africa*, April 3, 2013.

719. "Gazprom Operations in Niger," available from *wiki.openoil.net*.

720. "International Petroleum Snags 4 Blocks in Niger," *Petroleum Africa*, December 6, 2012. Subsequent press reports indicated that the names of International Petroleum's 4 blocks are Manga 1& 2, Aborak, and Tenere Ouest.

721. "Brandenburg Energy Updates on Due Diligence in Niger," *Reuters*, November 22, 1012.

722. "Niger to Gain Access to Chad-Cameroon Pipeline," *Petroleum Africa*, July 3, 2012.

723. Part of NNPC's inability to fund is cash obligations to IOCs is due to rampant corruption inside of the NOC. See Amrita Sen, "Nigeria: A New Dawn?" Oxford Energy Forum, Oxford, UK: November 2012, p. 20.

724. *Ibid.*

725. "With $3.5 billion, ConocoPhillips Cashes Out of Algeria, Nigeria," *Africa Oil and Gas Report*, January 2, 2013.

726. "Chevron Nigeria Joins the Divestment Train," *Africa Oil & Gas Report*, May 30, 2013.

727. The IEA reported that the Nigerian Government loses $7 billion annually to oil theft and illegal refining, mostly via pipeline damage from sabotage. See *www.nigeriaoilandgasintelligence.com*, November 24, 2102.

728. "Nigeria," EIA Country profile, October 16, 2012.

729. Due in part to MEND attacks, Nigeria's production fell to 1.7 million bpd in 2009, but climbed back over 2 million bpd in 2011 after some companies were able to repair damaged infrastructure. See "Nigeria: A New Dawn?" p. 20; and Helen Castell, "Nigeria Faces Return to the Bad Old Days," *AfrOil*, April 9, 2013.

730. "Nigeria," EIA Country profile, October 16, 2012.

731. 'Nigeria Fact Sheet," April 2012, available from *www.chevron.com*.

732. This figure is for crude oil. Total, for example, produced in 2011 a total of 287,000 BOE, which includes gas and gas condensates. See "Nigeria," available from *www.total.com*.

733. Philip Mobbs, "The Mineral Industry of Equatorial Guinea," Equatorial Guinea — 2011 [Advance Release], *U.S. Geological Survey Minerals Handbook — 2011*, p. 33.2.

734. "China Inc. Buys Another 28,000 BOPD of Nigeria's Production," January 2, 2013, available from *AfricaOilGasReport.com*.

735. "With 181,000 BOPD (net), China Becomes the Firth Largest Produce in Nigeria," *AfrOil*, April 9, 2013.

736. "CNPC in Nigeria," available from *www.cnpc.com.cn/en/cnpcworlwide/niger/*.

737. "Nigeria," available from *ongcvidesh.com/Assets.aspx?tab=2*.

738. "JVs/PSCs/Alliances," available from *oil-india.com/JVs.aspx*.

739. "Nigeria's Top Twenty Indigenous Crude Oil Producing Companies," *Africa Oil & Gas Report*, Table, 2012.

740. "About Mart," available from *www.martresources.com*.

741. EIA Short-term Energy Outlook (STEO), provided by US-AFRICOM, April 26. 2013.

742. "IOCs: 4 million bpd by 2020 impossible under PIB," *Vanguard*, April 4, 2013, as quoted in *www.vanguardngr. com/2013/04/4mbpd-by-2020-not-feasible-under-pib-iocs/*.

743. Moses Aremu, "Nigeria: Oil Production Could Fall to 1.84 MMBOPD," *Africa Oil & Gas Report*, May 2012.

744. "Nigeria," EIA Country profile, October 16, 2012.

745. "Oil, Gas and Security in Nigeria," available from *think-securityafrica.org/oilgas/oil-and-gas-in-Nigeria*.

746. "Nigeria," EIA Country profile, October 16, 2012.

747. "Petrole: Du petrole dans le lac Tchad?" ("Oil: Any Oil in Lake Chad?"), September 11, 2012, available from *www. zonebourse.com/LONDON-BRENT-OIL-4948/actualite/Petrole--Du-petrole-dans-le-lac-Tchad-14702549/*.

748. "JVs/PSCs/Alliances," available from *www.oil-india.com*.

749. "Oil, Gas and Security in Rwanda," available from *www. thinksecurityafrica.org*.

750. "Vanoil Receives Extension in Rwanda," *Petroleum Africa*, February 1, 2013.

751. See *www.vanoil.ca/s/Rwanda.asp*.

752. *U.S. Geological Survey Minerals Yearbook 2010* (Advance Release), p. 34.2.

753. "Kivuwatt," available from *www.contourglobal.com/portfolio/?id=11*.

754. "Uganda/Rwanda Pipeline Seeks Spark of Interest," *Petroleum Africa*, August 26, 2011.

755. *U.S. Geological Survey Minerals Yearbook – 2005*, p. 4.2.

756. "Commercial Value of Hydrocarbons Found in the Nigeria/Sao Tome Joint Area is Still Unknown," August 22, 2012, available from *www.macauhub.com.mo/en/2012/08/22/commercial-value-of-hydrocarbons-found-in-the-nigeriasao-tome-joint-area-is-still-unknown/*.

757. "AFREN Plc: Operations: Nigeria – Sao Tome and Principe JDZ: JDZ Block 1," January 13, 2013, available from *www.afren.com.*

758. See *investorshub.advfn.com/boards/hubstocks.aspx,* May 15, 2012.

759. "AFREN Plc: Operations: Nigeria – Sao Tome and Principe JDZ: JDZ Block 1."

760. "Block-2, Nigeria-Sao Tome & Principe, JDZ," available from *www.ongcvidesh.com/Assets.aspx?tab=2.*

761. "Sinoangol To Being Negotiations for oil block," *African Energy*, May 2, 2013.

762. Favour Nnabugwu, "Nigeria's Oranto Petroleum, 3 Others Grab Sao Tome Oil Blocks," October 24, 2011, available from *www.vanguardngr.com/2011/10/nigeria%e2%80%99s-oranto-petroleum-3-others-grab-sao-tome-oil-blocks/*.

763. "Benin, Burkina Faso, and Sao Tome E Principe – 2010 (Advance Release)," *U.S. Geological Survey Minerals Yearbook – 2010*, p. 4.3.

764. See *www.equatorexploration.com/operations.*

765. See *erhc.com.*

766. "Doing Business in Sao Tome and Principe," *2012 Country Commercial Guide*, U.S. Foreign Commercial Service, Chap. 1.

767. "Sao Tome annonce son port pétrolier en eaux profondes" ("Sao Tome Announces Its Deep-Water Petroleum Port"), Agence Ecofin, February 28, 2012.

768. Africa Fortesa Corporation, available from *www.fortesa.com*.

769. "Senegal—Overview of Oil and Gas Exploration and Main Players," June 2011, available from *mergersandacquisition reviewcom.blogspot.com/2011/06/senegal-overview-of-oil-and-gas.html*.

770. *U.S. Geological Survey Minerals Yearbook—2010*, Advance Release, p. 18.4.

771. See *www.tullowoil.com/*.

772. "Cairn Energy Unveils Deal in Senegal . . ." available from *www.theguardian.com/business/marketforceslive/2013/mar/19/cairn-energy-senegal-greenland*.

773. See *www.far.com.au/senegal/*.

774. "Elenilto Lands PSC in Senegal," *Petroleum Africa*, January 2, 2013.

775. See *www.elenilto.com/projects/oil-gas/senegal/*.

776. Even in English, the JDZ is often referred to by its French acronym AGC, or Agence de Gestion et de Cooperation, which actually refers to the Joint Commission that manages the JDZ.

777. See *www.ophir-energy.com/about-us/who-we-are.aspx*.

778. "Oil, Gas and Security in Senegal," available from *www.thinksecurityafrica.org*.

779. "Seychelles Sees its First 3D Shoot," *Petroleum Africa*, December 13, 2012.

780. "Seychelles MC2D Ready for Processing," *Petroleum Africa*, July 20, 2011.

781. "Offshore Seychelles a 'Flagship' Project," available from *www.whlenergy.com/irm/content/seychelles.aspx?OriginalCategoryId=220*.

782. "Seychelles Delays New Block Licensing," *Petroleum Africa*, November 5, 2012.

783. George Thande, "Seychelles To Change Oil Laws, Invite Exploration Bids," *Reuters*, March 13, 2013.

784. "Seychelles to Revamp PSAs," *Petroleum Africa*, March 15, 2013.

785. *U.S. Geological Survey, Minerals Yearbook 2011*.

786. "WHL Delivers on 3 Initiatives in Seychelles," *Petroleum Africa*, October 11, 2012.

787. "Resource Report on Seychelles Potential," *Petroleum Africa*, November 11, 2011.

788. "WHL Energy Closes In on Seychelles Farm-In Partner," *AfrOil*, April 2, 2013.

789. "Seychelles Sees its First 3D Shoot."

790. "Seismic Bulks Up Afren's East African Resource Portfolio," *Petroleum Africa*, August 23, 2013.

791. See *www.afren.com/operations/seychelles/*; "Frontier Exploration," *GEO ExPro*, May 2006, p. 42.

792. See *www.afren.com/operations/seychelles/*.

793. "Frontier Exploration," *GEO ExPro*, May 2006, p. 42.

794. "Jupiter-1 Hits Oil Offshore Sierra Leone," *Petroleum Africa*, February 22, 2012; Sierra Leone, available from *www.petroleumafrica.com/jupiter-1-hits-oil-offshore-sierra-leone/*.

795. See *www/africanpetroleum.com.au/our-projects/sierra-leone*.

796. *U.S. Geological Survey Minerals Yearbook – 2011*, Advance Release, p. 35.5; "Oil exploration in Sierra Leone in 2011 . . . Maybe a New Frontier," June 14, 2011, available from *mergersand acquisitionreviewcom.blogspot.com/2011/06/sierra-leone-war-poverty-and-oil.html*.

797. See *www.elenilto.com/projects/oil-gas/sierra-leone*.

798. "Chevron Awarded Deepwater Interest Offshore Sierra Leone," September 26, 2012, available from *www.chevron.com*; "Sierra Leone Teamed Various Oil Block Bidders Together for Four of the Block, Telling Them to Move Forward as Partners in Negotiation," August 21, 2012, available from *www.thisissierraleone.com*.

799. See *www.thisissierraleone.com/talisman-energy-sells-down-block-sl-4b-10-in-sierra-leone/*, November 8, 2012.

800. "Jupiter-1 Hits Oil Offshore Sierra Leone," *Petroleum Africa*, February 22, 2012, available from *www.repsol.com/es en/*. See also *www.repsol.com/es es/corporacion/conocer-repsol/repsol-en-el-mundo/sierra-leona.aspx*.

801. "Sierra Leone to See $100 million from Lukoil." *Petroleum Africa*, October 21, 2011; and "Projects – SL-5-11 Block," available from *www.panatlanticexploration.com*.

802. "Lukoil and Vanco Land in Sierra Leone," *Petroleum Africa*, July 22, 2011.

803. See *www.repsol.com/es es/corporacion/conocer-repsol/repsol-en-el-mundo/tunez.aspx*.

804. Angelia Sanders *et al.*, "Emerging Energy Resources in East Africa," Norfolk, VA: Civil-Military Fusion Center Mediterranean Basis Team, September 2012, p. 3.

805. "Oil, Gas and Security in Somalia," available from *www. thinksecurityafrica.org*.

806. See *www.mbendi.com/land/af/so/p0005.htm*.

807. Sanders *et al.*, "Emerging Energy Resources in East Africa," p. 3.

808. William Macpherson, "New Somali Government Looks to Revive Oil Search," *African Energy*, May 16, 2013.

809. "Somaliland's Door is Open for Investors," November 2, 2012, available from *Somalilandpress.com*.

810. "Somalia/Kenya Maritime Dispute Heating Up," *Petroleum Africa*, July 10, 2012.

811. "Competent Person's Report of Certain Properties in Puntland, Somalia as of 31st December 2010," Alton, UK: Gaffney, Cline & Associates, May 2011, p. 8.

812. See *en.wikipedia.org/wiki/Energy in Somalia#Energy*.

813. "Operations Somalia (Puntland)," available from *www.www.africaoilcorp.com/s/current-activities.asp*.

814. "Somalia Spuds First Well in Decades," *Petroleum Africa*, January 17, 2012.

815. "Oil, Gas and Security in Somalia," available from *www.thinksecurityafrica.org*.

816. "Range To Merge with Timis Company," *African Energy*, May 2, 2013.

817. See *www.hornpetroleum.com/s/Dharoor Nugaal Block.asp?ReportID=499459& Type=Dharoor-and-Nuggal-Blocks& Title=Block-Summaries*.

818. "Horn Petroleum Third Quarter 2012 Results," Horn Petroleum news release, November 20, 2012.

819."Puntland Drilling Reveals Active System," *Petroleum Africa*, April 13, 2012.

820. See *www.redemperorresources.com/irm/content/puntland.aspx?RID=216*.

821. "What Puntland Success Could Mean," *Petroleum Africa*, February 8, 2012.

822. "Somalia (Puntland)," available from *www.africaoilcorp. com/s/current-activities.asp*.

823. "Shabeel North Plugged in Puntland," *Petroleum Africa*, August 28, 2012.

824. Sanders *et al.*, "Emerging Energy Resources in East Africa," p. 3.

825. "Somaliland: Genel Energy to Farm-In to Jacka Resources Odewayne Block, Somaliland," November 12, 2012, available from *www.energy-pedia.com/news/ somaliland/new-152401*. and "Odewayne Block Overview," November 14, 2012, available from *www.jackaresources.com.au*.

826. "Somaliland," available from *www.genelenergy.com/ operations/somaliland.aspx*.

827. "Jacka Moves Toward Seismic in Somaliland," *Petroleum Africa*, March 7, 2013.

828. See *www.ophir-energy.com/our-assets.aspx*.

829. "Somalia: DNO International Takes Somaliland Block," *African Energy*, May 2, 2013.

830. "Oil, Gas and Security in South Africa," available from *www.thinksecurityafrica.org*.

831. "Petroleum Exploration and Production Activities in South Africa," available from *www.impactoilandgas.co.uk*.

832. "ExxonMobil to Explore Offshore South Africa," December 17, 2012, available from *news.exxonmobil.com*.

833. "Exxon Extends Africa's Energy Enterprise," UPI, December 21, 2012.

834. See *www.impactoilandgas.co.uk.*

835. See *www.petrosa.co.za/products and services/Pages/Crude. aspx.*

836. *Country Report on South Africa,* Washington, DC: U.S. Energy Information Administration, updated January 17, 2013.

837. "Sunbird Executes Agreement to Acquire 870 Bcf Ibhubesi Gas Project Offshore South Africa," Sunbird Energy ASX Announcement, December 21, 2013.

838. "Exxon Extends Africa's Energy Enterprise."

839. "PetroSA and Cairn India Sign Agreement for Block 1 in South Africa," PetroSA/Cairn joint press release, August 16, 2012.

840. "Sunbird Executes Agreement to Acquire 870 Bcf Ibhubesi Gas Project Offshore South Africa."

841. "Thombo Launches South Africa Shoot," *The Petroleum Professional,* January 29, 2013.

842. See *www.afren.com/operations/south africa/block 2b/.*

843. "Falcon and Chevron Join Forces in Karoo Basin," *Petroleum Africa,* December 13, 2012.

844. See *www.falconoilandgas.com/karoo-basin-sa.*

845. "Chevron Joint Shale Hunt in South Africa," *The Wall Street Journal,* December 14, 2012.

846. See *www.shell.com/zaf/aboutshell/shell-businesses/e-and-p/ karoo/karoo.html.*

847. "Kinetiko Updates its Amersfoort Ops," *Petroleum Africa,* February 21, 2013.

848. See *kinetiko.com.au/projects/amersfoort-project.html.*

849. *Country Report on South Africa,* EIA.

850. "Sudan and South Sudan," Country Analysis Briefs, Washington, DC: U.S. Energy Information Administration, updated March 19, 2012.

851. "Resuming South Sudan Exports Hits Snag," *Petroleum Africa*, November 29, 2012.

852. "South Sudan to Export Crude Oil by Road Through Ethiopia," *Pan-African News*, March 2, 2013.

853. Sanders *et al.*, "Emerging Energy Resources in East Africa," p. 3.

854. "Oil, Gas and Security in Sudan and South Sudan," available from *www.thinksecurityafrica.org*.

855. Jill Shankleman, "Oil and State Building in South Sudan — New Country, Old Industry," Washington, DC: United States Institute of Peace, July 2011, p. 1.

856. "Pipeline in the Works for South Sudan," *Petroleum Africa*, February 21, 2013.

857. Kelly Gilblom, "South Sudan Says Oil Pipeline via Kenya to Cost $3 Billion," *Reuters*, August 10, 2012.

858. "South Sudan in Ethiopia-Djibouti Oil Pipeline Deal," BBC News, February 9, 2012; Jared Ferrie, "China to Loan South Sudan $8 Billion for Infrastructure Projects," Bloomberg, April 28, 2012.

859. Bill Farren-Price, "The Sudanese Struggle for Stability: Long-Term Energy Security Hinges on Deeper Bilateral Political Progress," Oxford Energy Forum, Oxford, UK, November 2012, p. 24.

860. "South Sudan and Sudan Reach Oil Flow Agreement," *Petroleum Africa*, March 13, 2013.

861. EIA Short-term Energy Outlook (STEO), provided to author by USAFRICOM source on April 25, 2013.

862. "Sudan Signs Oil, Gas Exploration Deal in North East," *Reuters*, August 19, 2010; and "Khartoum," available from *www. fennocaledonian.com*.

863. "Sudan and South Sudan," Country Analysis Briefs, EIA.

864. "Sudan and Brazil to Cooperate," *Petroleum Africa*, November 27, 2012.

865. "Sudan Adds to Production," *Petroleum Africa*, December 28, 2012.

866. See *en.wikipedia.org/wiki/South Kordofan*.

867. Note: The South Sudan assets of Sudapet were transferred to Nilepet, the new South Sudan NOC, per "South Sudan," available from *www.ongcvidesh.com/Assets.aspx?tab=2*.

868. "Sudan and South Sudan," Country Analysis Briefs, EIA; "Exploration and Production—South Sudan," available from *www.trioceanenergy.com*.

869. "ONGC Videsh Limited,"available from *www.ongcindia. com*.

870. "Upstream Africa Acreage—A New Hub for the Group in East Africa," *Total Factbook 2011*.

871. Jen Alic, "Desperate South Sudan Carves Up Total Concession," November 8, 2012, available from *www.oilprice.com*.

872. Jenny Gross, "South Sudan Offers Total Block to Exxon, Kuwait Petroleum," September 14, 2013, available from *www. rigzone.com*.

873. "Total to Bring ExxonMobil and Kufpec into South Sudan," *Petroleum Africa*, June 4, 2013.

874. "Tanzania Gas Bounty Leads to New Regulations," *Petroleum Africa*, June 20, 2012. Note: This 34 tcf figure comes from the 28.74 tcf figure cited in this article, plus the 5 tcf in new discoveries announced by Statoil in December 2012 and February 2013.

875. "2012 Tanzania Country Commercial Guide," Washington, DC: U.S. Foreign and Commercial Service (USFCS).

876. "Ophir Energy Reveals Jodari Field Discoveries," December 14, 2012, available from *www.bg-group.com/288/where-we-work/tanzania/*. "Papa-1 — Play Opening First Cretaceous Gas Discovery Outboard of the Rufiji Delta, Block 3, Tanzania," Ophir Energy PLC, press release, August 2, 2012.

877. "Tanzania: Commercializing Gas," Ophir presentation to Investors, 2012.

878. "Ophir Lines up New Drilling in Tanzania," *Petroleum Africa*, April 11, 2013.

879. "BF Completes Tanzania's Mzia Well Test," *Petroleum Africa*, May 2, 2013.

880. "Ophir Raises Block 1 Reserves Estimate," *African Energy*, May 16, 2013.

881. Gismatullin, "Kenya's Deepwater Debut Heralds East Africa's First Oil: Energy."

882. Eduard Gismatullin and Brian Swint, "Statoil $1.2 billion Tanzania Find May Hold Oil in New Play," Bloomberg, February 24, 2013; "New Discovery for Statoil and ExxonMobil Off Tanzania," *Petroleum Africa*, December 24, 2012; "Statoil and ExxonMobil Knock Out Tanzania Discovery," *Petroleum Africa*, March 19, 2013.

883. "BG to Present LNG Terminal to Tanzanian Officials," *Petroleum Africa*, May 3, 2013.

884. "Rovuma PSA," available from *www.solooil.co.uk*. See also *www.aminex-plc.com*; and "Aminex Continues Farm Out Discussions in Tanzania," *Petroleum Africa*, March 22, 2013.

885. Tanzania to Increase Royalties and Fees," *Petroleum Africa*, July 30, 2012.

886. *Ibid.*

887. "Tanzania Vows No Contracts to Be Revoked," *Petroleum Africa*, September 20, 2012.

888. Sa'd Bashir, "East Africa on Sale," *Africa Oil & Gas Report*, May 21, 2013.

889. "Tanzania to Offer New Deepwater Acreage," *Petroleum Africa*, July 16, 2012.

890. "Afren Tangos in Tanzania's Tango Block," *Petroleum Africa*, March 25, 2011.

891. "Tanzania," available from *www.maureletprom.fr*.

892. "Kilosa-Kilombero Licence," "Pangani Licence," available from *www.swala-energy.com*.

893. "Tanzania," available from *nuenergygas.com/projects/eastern-africa/*.

894. "Tanzania to Award Acreage to Total," *Petroleum Africa*, February 14, 2013.

895. "Jacka Acquires Ruhuhu Block in Tanzania," *Petroleum Africa*, March 21, 2013.

896. *U.S. Geological Survey Minerals Yearbook – 2010*, p. 39.4.

897. "Tanzania Sees Orca Spud," *Petroleum Africa*, February 8, 2012.

898. Maurel et Prom's website indicates that it also has a 48.06 percent ownership share in a production permit for Mnazi Bay. "Tanzania," available from *www.maureletprom.fr*.

899. "Large-scale Gas Monetization Projects," available from *www.wentworthresources.com/large-scale.php*.

900. "$1.225 bn Deal for Gas Pipeline Inked," June 23, 2012, available from *www.tanzanianews24.com*.

901. "ENI Confirme l"Existence de Petrole au Large du Togo" ("ENI Confirms Existence of Oil Along Togo Coast"), Agence Ecofin, June 9, 2012.

902. "Le Groupe ENI se prepare a forer un second bloc pour le petrole offshore Togolais" ("Togo : ENI Group Prepares to Drill Second Well for Offshore Togolese Oil"), *Xinhua*, January 3, 2012; "Togo Asset Overview," DGS Energy Oil & Gas.

903. "Mauritania, Benin and Togo: Overview of Oil and Gas Discoveries and Exploration in 2011," available from *mergersandacquisitionreviewcom.blogspot.com/2011/06/mauritania-benin-and-togo-overview-of.html*.

904. "ENI Discovers Oil in Tano Basin Offshore Ghana," available from *www.epmag.com*.

905. "Tunisian Oil and Gas," available from *deplc.com/tunisian-oil-and-gas*.

906. "Regions Petroliferes en Afrique" ("Petroleum Regions in Africa"), available from *fr.wikipedia.org*.

907. "Tunisia," *U.S. Geological Survey Minerals Yearbook — 2010*, p. 40.3.

908. "Repsol in Tunisia," available from *www.repsol.com*; "Tunisia," available from *www.chinookenergyinc.com*; "Our Operations," available from *www.cooperenergy.com.au/*; "Assets in Tunisia," available from *www.dno.no*; "Tunisia," available from *www.paresources.com*; "Latest News," available from *www.sonderesources.com/*; "Chorbane Permit—Onshore Tunisia (70%)," available from *www.gulfsands.com/s/Tunisiaasp*; "Tunisia," Vietnam National Oil and Gas Group, available from *english.pvn.vn*.

909. "Tunisia," available from *www.mbendi.com/land/af/tu/p0005.htm*.

910. "Tunisian Oil and Gas," available from *deplc.com/tunisian-oil-and-gas*.

911. "Tunisia," available from *www.bg-group.com/237/where-we-work/tunisia/*.

912. "Tunisian Premium Welcome at Chinook," *AfrOil*, April 2, 2103.

913. "Uganda—Disposal of the Interests in Blocks 1 and 3A," available from *www.heritageoilplc.com/about-us/history.aspx*.

914. "Uganda," available from *www.tullowoil.com*.

915. "Tullow Files Suit Against Uganda," *Petroleum Africa*, December 18, 2012.

916. "The Business Case for Large Scale Uganda Refinery, *Africa Oil and Gas Report*, #9, December 2012.

917. "Museveni Set on a Refinery for Uganda," *Petroleum Africa*, December 24, 2012.

918. "Uganda Reaches Agreement on Refinery," *Petroleum Africa*, April 17, 2013.

919. "Uganda MOU on the Way for Lake Albert Partners," *Petroleum Africa*, May 9, 2013.

920. "Uganda: Export Pipeline is the Bankable Asset," *Africa Oil and Gas Report*, October 16, 2012.

921. Tabu Butagira, "Museveni Strikes Deal on Oil and Arms with Russia," *Africa Review*, December 12, 2012.

922. "Analysis: New Law Fails to East Oil Concerns in Uganda," Nairobi, Kenya: Integrated Regional Information Network (IRIN), December 13, 2012.

923. "Uganda Oil and Gas Exploration in 2011 & 2012," July 15, 2011, available from *mergersandacquisitionreviewcom.blogspot.com/2011/07/uganda-oil-and-gas-exploration-in-2011.html*.

924. "13 Blocks to be Offered by Uganda," *Petroleum Africa*, January 25, 2013.

925. "Uganda," available from *www.tullowoil.com*.

926. Sanders *et al.*, "Emerging Energy Resources in East Africa," September 2012, p. 4.

927. Uganda country report, *U.S. Geological Survey Minerals Yearbook — 2011*, p. 42.2.

928. Eduardo Pereira and Elison Karuhanga, "Brazil and Uganda: Government Intervention and Oil Development Prospects," Oxford Energy Forum, Oxford, UK, November 2012, p. 18.

929. "Uganda's Reserves Rise but Development Stalls," *Petroleum Africa*, September 18, 2012.

930. "Uganda Appoints Panel for Oil Development," *Petroleum Africa*, July 9, 2012.

931. "Oil & Gas — Ghana and Uganda — The Quick and the Slow," November 18, 2012, available from *www.thefreelibrary.com/Ghana+and+Uganda--the+quick+and+the+slow.-a0309793722*.

932. EIA Short-Term Energy Outlook (STEO), as provided to author by source at USAFRICOM on April 25, 2013.

933. Ben Shepherd, "Oil in Uganda — International Lessons for Success," London, UK: Chatham House, February 2013.

934. "Morocco and Western Sahara 2010 (Advance Release)," New Cumberland, PA: *U.S. Geological Survey*, p. 30.5.

935. "Has Total Re-entered the Disputed Western Sahara?" *Petroleum Africa*, November 30, 2012.

936. "Western Sahara Northwest Africa," available from *www.wessexexploration.com*.

937. "Saharawi Arab Democratic Republic (SADR)," available from *www.ophir-energy.com/about-us/who-we-are.aspx*.

938. "Premier Buys into West African Assets," press release, May 28, 2003, available from *www.premier-oil.com/premieroil/media/press/premier-buys-into-west-african-assets*.

939. "Oil, Gas and Security in Zambia," available from *thinksecurityafrica.org/oilgas/oil-and-gas-in-zambia/*.

940. Maxwell M.B. Mwale, Statement before Zambia's Parliament on the "Status of Petroleum Exploration in Zambia," Minister of Mines and Mineral Development, 2010, available from *www.parliament.gov.zm*.

941. "Zambia," available from *www.petrodel.com*.

942. Mwale.

943. "Frontier Resources Scores Zambia Block," *Petroleum Africa*, April 14, 2011. Note: Another source, the *Lusaka Times*, indicated that the Zambian Government issued 11 licenses in 2011, not three ("State to Revoke License for Oil, Gas Prospectors," September 21, 2012, available from *www.lusakatimes.com*).

944. "Oil and Gas potential in Zambia, Malawi, and Botswana," *African Energy*, May 16, 2013.

945. "State to Revoke License for Oil, Gas Prospectors."

946. "Zambia Says Use It or Lose It," *Petroleum Africa*, February 5, 2013.

947. "Projects: Zambia," available from *www.friplc.com/pdf/projects/gravity-survey-zambia.pdf*.

948. "Frontier Progresses Zambia Exploration," *Petroleum Africa*, September 7, 2011.

949. "Oil, Gas and Security in Zimbabwe," available from *thinksecurityafrica.org/oilgas/oil-and-gas-in-zimbabwe/*.

950. "Zimbabwe Exploring for Oil and Gas in Zambezi Valley — Mpofu," December 27, 2011, available from *www.bulawayo24.com*.

951. *Ibid.*

APPENDIX II

DEFINITIONS OF OIL AND GAS "RESERVES" AND QUALITIES OF CRUDE OIL[1]

Reserves.

The most widely used definitions today of reserves are from the Petroleum Resources Management System of the American Society of Petroleum Engineers (SPE). According to SPE Guidelines, "reserves" are a subset of "resources," representing the part of resources that are commercially recoverable and have been justified for development. Reserves can be subsequently divided into the following three categories, depending on certainty of recovery:

1. **Proved Reserves.** The highest valued category of reserves is "proved" reserves. Proved reserves have a "reasonable certainty" of being recovered, which means a high degree of confidence that volumes will be recovered. Reserves must have all commercial aspects addressed.

2. **"Probable" or "Possible" Reserves**. A lower category of reserves, commonly combined and referred to as "unproved reserves," with decreasing levels of technical certainty. Probable reserves are volumes that are defined as less likely to be recovered than proved, but more certain to be recovered than "possible reserves." Possible reserves are reserves which analysis of geological and engineering data suggests are less likely to be recoverable than probable reserves. **"2P"** is used to denote the sum of proved and probable reserves. The best estimate of recovery from committed projects is generally considered to be the 2P sum of proved and probable reserves.

3. **Resources.** Denotes less certainty than "reserves" because some significant commercial or technical hurdle must be overcome prior to there being confidence in the eventual production of the volumes. A sub-category is **contingent resources,** which are potentially recoverable but not yet considered mature enough for commercial development due to technological or business hurdles. Another sub-category is **prospective resources.** Estimated volumes associated with undiscovered accumulations. These represent quantities of petroleum that are estimated, as of a given date, to be potentially recoverable from oil and gas deposits identified on the basis of indirect evidence but which have not yet been drilled. This class represents a higher risk than contingent resources since the risk of discovery is also added.

Crude Oil Qualities.

1. **Density.** Oil density is generally expressed in degrees using an API scale. This is a specific gravity scale developed by the American Petroleum Institute (API), designed to measure the relative density of various petroleum liquids. The measure is expressed in degrees and most values fall between 10° and 70° API gravity. The specific gravity of oil is its relative density to water at 60° Fahrenheit.

2. **Light oil.** Otherwise known as "conventional oil," or "sweet crude," light oil has an API gravity of 22° or over. For example, the oil produced from Libyan fields is typically very "light," and the country's nine export grades have API gravities that range from 26-43.

3. **Heavy oil.** Dense, viscous oil with low API gravity. Definitions vary, but it is generally accepted that

the upper limit for heavy oils is 22° API. In Venezuela for example, the Bachaquero heavy crude oil has an API gravity of 17°. Heavy oils are usually not recoverable in their natural state through a well or by using ordinary production methods. Most need to be heated or diluted so that they can flow into a well or through a pipeline.

4. **Extra heavy oil.** Extra heavy oil has an API gravity of less than 10°.

5. **Bitumen.** Otherwise known as "oil sands," bitumen shares many attributes of heavy oil, but is even more dense and viscous.

ENDNOTE - APPENDIX II

1. "Petroleum Resources Management System," World Petroleum Council, undated paper, 49 pgs.

U.S. ARMY WAR COLLEGE

Major General Anthony A. Cucolo III
Commandant

*****

STRATEGIC STUDIES INSTITUTE
and
U.S. ARMY WAR COLLEGE PRESS

Director
Professor Douglas C. Lovelace, Jr.

Director of Research
Dr. Steven K. Metz

Author
David E. Brown

Editor for Production
Dr. James G. Pierce

Publications Assistant
Ms. Rita A. Rummel

*****

Composition
Mrs. Jennifer E. Nevil

www.ingramcontent.com/pod-product-compliance
Lightning Source LLC
Chambersburg PA
CBHW082351270326
41935CB00013B/1580